Clouds Above

Plausible Science Fiction

SHEPHERD KING PUBLISHING, LLC

This is a work of fiction. Names, characters, businesses, places, events, locales, and incidents are either the products of the author's imagination or used in a fictitious manner. Any resemblance to actual persons, living or dead, or actual events is purely coincidental.

Clouds Above / Thompson —1st ed.

ISBN:978-0-9976556-0-5

Library of Congress Control Number: 2020925552

Print On Demand (POD) copies available in hardback and paperback.

SHEPHERD KING PUBLISHING, LLC

Author website: **michaelthompsonauthor.com**

This book was set in Garamond, 12 pt.

Printed in the United States of America.

10 9 8 7 6 5 4 3 2 1

For Tempe, my wife, my life, my love

ACKNOWLEDGEMENTS

With special thanks to Twyla Dixon, my editor who made it her goal to correct my wrongs and make everything perfect, including being a joy to work with.

With research thanks to Maude Barlow, author of *Blue Covenant*, The New Press, 2007, on the impending global water crisis. I've received permission and cited her work where appropriate. If you want to seriously research the facts about our planet's coming crisis, you'll want to read her prophetic warnings on the subject.

Thanks also to Robbyn Weaver and Ward Archer, for reading an early version and offering valuable input.

With creative design thanks to Disciple Design for the cover and illustrations; and Nancy Roe for book formatting.

With unique thanks to Google Earth for making it possible to find and explore the villages of Kupup and Darjeeling in India, and Gedu and Chhukha in Bhutan.

AUTHOR'S NOTE

I have a confession. I enjoy writing in first person singular. I've done that with my heroine, Grayson Fields, but only in the chapters in which she appears. All other chapters are written in third person.

Regarding my use of the term *plausible science fiction*, this story is not only plausible, but predictive. Hopefully, the entertainment you receive from the story is only a taste of what you gain in new knowledge about our future on this planet.

पानी = paanee = water in Hindi.

MAIN CHARACTERS

in order of appearance

Grayson Fields, Ph.D., scientist
Jesse, chauffeur
Julius Schwarzkopf, Ph.D., scientist
Senator Bill Bennett, New Jersey
Steve Muller, CEO, Purity Worldwide
Bernard Loren, Ph.D., scientist, Purity Worldwide
Brooks Turnage, missionary
Rinku Patel, villager
Rashman, villager
Jaiman, villager
Pete Gerritsen, NOAA
Senator Ray Gill, Arizona
Senator Margaret Chambers, New York
Latoya, pilot

– 1 –

D.C.

Wisconsin Ave., Washington, D.C. October 2035, 4 p.m.

"DR. FIELDS, ARE YOU CERTAIN YOU know what you're doing?"

"No, I'm not, Jesse. But I should only be a minute, hopefully."

Again, I glanced out the window at the family in the field and checked my watch for the outside temperature. 96°.

I stuffed four large bottles of water in a bag and stepped out onto the shoulder of a hot, bustling Wisconsin Avenue.

A boy and a girl played catch with a ball in a vacant patch of land on the grounds of the Washington National Cathedral. A man and a woman sat on shredded nylon chairs in front of their makeshift tin and cardboard home. The man, middle age; she, probably in her 30s, very pregnant.

The illegality of squatters had never bothered me. Quite the opposite. I was sensitive to their plight.

As I approached, the woman asked, "Are you the one to make it rain?" her eyes dull and vacant.

"She's crazy. Not herself anymore," the man said.

I held out the bottles, one by one. The man spoke again. "I was rich once. I had money once."

"I'm sorry."

I wanted to cry for them.

"Are you the one to make it rain?" she asked again.

"I don't think so."

After they accepted the last bottle I turned and hustled back to the car.

Jesse asked, "Dr. Fields, did you accomplish your mission?"

"I did, Jesse," I said, smiling. "And thank you for pulling over in the middle of this traffic."

"You are most welcome. Where to next, Dr. Fields?"

"Home."

"Very well." He eased the Tesla-X back onto Wisconsin.

I felt good about helping one family, but I was after far more. I wanted to save billions. I wanted to save our planet.

Tired, sweaty, I leaned back on the headrest and soaked up the dry scenery passing by at seventy miles an hour.

I was saddened by this ugly scene in our nation's capital. Dust covered everything—the streets, cars, homes, everything. Every bush, every leaf of vegetation was shriveled, dying for lack of water.

After a few minutes Jesse turned right onto Nebraska, then later, left onto Connecticut Avenue NW.

It wasn't long before he slowed and stopped a few yards from a blue vehicle blocking the street.

Through the filthy windshield ahead I could see the problem—a Purity Worldwide water truck. Two uniformed men in SWAT gear, carrying space age weapons and wearing Purity Worldwide sleeve patches walked right in front of us.

They met two other men on the sidewalk carrying five-gallon jugs of water. The armed men kept an eye on the water carriers. Angry, yelling neighbors on the street approached the men.

"Can you hear them, Jesse?"

"I hear them clearly."

"What are they saying?"

"They are asking for the water."

Emotions were intense; people screamed at each other, arms flailing. Some waved cash hoping to get the water carriers' attention.

To my left, a woman stepped out onto her doorway where she took control of a five-gallon jug. She yelled something abrasive at the neighbors before retreating inside, slamming the door.

"Can we get out of here, Jesse?"

"Yes, ma'am." Suddenly we were easing around the truck and most of the sidewalk trash bags. We sped away, the tension now behind me.

"Thank you, Jesse."

"Most welcome, Dr. Fields."

A minute or two passed.

"How long has it been, Jesse?"

"For what, Dr. Fields?"

"Since the start of the drought."

"It's been three years, two months, and six days."

"And it's only going to get uglier, isn't it?"

"Yes, I would say so. My sources tell me—"

"No, Jesse. I don't want to hear it again. But thank you."

People in America were sick and tired of the new laws restricting water usage. Scheduled home water uses had become the bane of every family.

Grey water (shower, bath, and kitchen) was processed through home filtration systems. Even it, after recycling umpteen times, had a limited life.

In America we're mostly inconvenienced while most of the world is dealing with far worse—they're starving.

I had always wondered why the world's brain trusts never saw it coming. *We've become reactive to everything, not proactive. I suppose it happens to every bloody bloated civilization. All the signs were there. Our own arrogance blinded us to the truth. Why didn't we see it coming?*

An old adage about marriage popped into my head, one that epitomized the current drought and pretty much summed up my own brief marriage: "Civilizations and marriages don't end in one fatal blow to the heart; they end in a thousand tiny cuts."

Yes, I was married for two years. To the handsome Phillip Percy. And like so many others, we finally called it quits after a thousand tiny cuts. I swear to you he dished out more than I did. Thank goodness we never had children.

My life before Phillip Percy was one of study-study-study for my Ph.D. and never forget that *I'm fat and unattractive and not likely to ever find that someone special.*

Then, one day, a handsome young man named Phillip comes along. He tells me how perfect and beautiful I am.

Yep, I was the naive victim in need of self-confidence. I was a book worm getting my Ph.D. while he was making good money servicing clients for Seinfeld & White Investments. I heard he was also servicing the office girls.

Why he chose me I'll never know.

But the good news is, his thousand little cuts about my body turned me into a Manduu fitness freak. I got fit. He got fat.

Jesse stopped at a red light next to an Exxon. The prices were an ugly reminder.

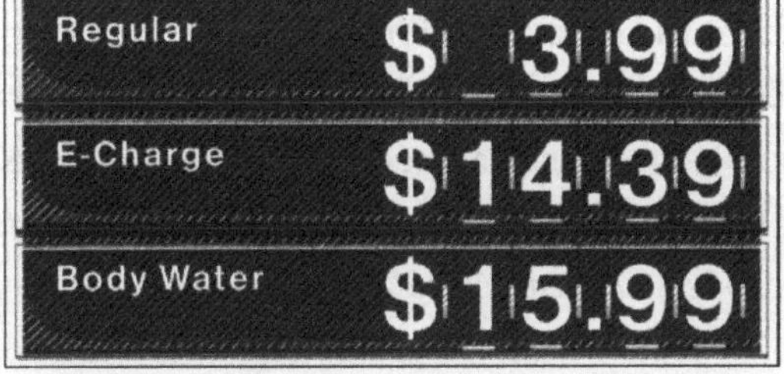

How could people even afford body water?

In the early 2000s it was called white water. But due to racial strifes of the early 20s, white water was changed to body water in 2026. More politically correct, congress thought. "Henceforth, pure drinking water shall no longer be referred to as white water. It shall be known as body water."

It was the law of the land. And I was in favor of it. Grey water and black water designations remained the same.

Ahead was one of Purity Worldwide's beautiful, aqua blue billboards with their non-stop, famous, blinking slogan: *Purest Body Water in the World.* They were plastered not only around Washington but all over the world in hundreds of languages and places much like every nook and cranny Coca-Cola used to occupy.

Still, I couldn't help but smirk, knowing what grandfather had told me about Purity Worldwide.

Minutes later, we were in Chevy Chase, turning into 5408 Highland, where we stopped at two 15-foot-high gates carrying more wrought iron than a French Quarter balcony. We waited and watched as the gates creaked open, rusty from neglect.

Once inside, Jesse and I rode over a winding, crushed shell driveway before circling an ornate but dry fountain. We stopped at the front of a stone and ivy-engulfed three-story mansion—more like a green castle with turrets. We were facing the home of my grandfather, Dr. Julius Schwarzkopf, the famous scientist I love and adore.

I headed to the front door while Jesse walked toward the five-car garage.

Even before I moved in with Grandfather, I loved to tease Jesse, just to see his reaction. On this day I dropped my chin and cast a coy smile in his direction. "You know Jesse, you and I, we should go out to dinner some time."

Jesse's stern face turned into a frown, "Stop that, Missy. You know better than to flirt with a man-made." He turned and headed for the garage, his gait slightly stymied, almost human-like.

Jesse had no last name. Bots didn't need a last name. Their original owners gave them only a first name in case they were sold or traded to another family. Neither did bots actually drive the cars. Since driverless cars never caught on, Tesla was the first to put the computers inside a bot.

Research concluded it made us feel more at ease having a man-made at the wheel. They quickly became personal chauffeurs.

Stepping into the dark foyer I was greeted by Arnold, a real flesh and blood human, who acted more like a robot than robots.

"Your grandfather is in the library, not to be disturbed. I believe he is with a Senator. They've been at it for nearly an hour now. So, best if you head up to your room and give him some space. I'm sure he'll join you for dinner at seven."

"Thank you, Arnold."

As I hit the first step and kept a fast pace to the top, I muttered, "It's time for Arnold to do whatever it is he does … so I can do my thing."

I tossed my backpack on the bed and tiptoed back downstairs.

Arnold was nowhere in sight. I eased up next to the massive oak doors so I could hear Grandfather and this Senator.

Grandfather's deep voice was nothing short of intimidating. "Well, I have to say, Bill, you have more intestinal fortitude to hang yourself out like this than I imagined. If you allow congress to pass this bill, it'll throw the undeveloped countries into holy hell. These people can't afford decent body water as it is! Bill, help me understand why you are so beholden to these water companies."

"I'm not. Look, Julius, there's something else. At our NOAA facility in the Himalayas, we're building an ocean cloud-seeding simulator like you can't imagine. It'll be twice the size of yours and be fully tech'd out with the latest of everything."

"I've heard."

"We want you to go there and conduct your experiments. What you have here is ancient. You'll have a state-of-the-art simulator, Julius."

"Yes, that part intrigues me, I admit. But I can find the answer with my own simulator. I am close to discovering the algorithm."

"So you've told me. Going on what, two years now?" A long sigh came from the Senator. "Look, Julius, in my case, I don't have a choice. I have to go with the movement, and right now it's with this new legislation. C'mon, Julius, NOAA is begging you to go to India."

"I wasn't aware of any begging. Hell, I haven't even heard from NOAA."

"That's why I'm here. I'm their liaison on this. Trust me, you'll be able to solve the problem much faster."

"Are the water companies involved?"

"They're helping, yes, to fund it. That's all, Julius. Congress is not about to fund any more for the project. They think it's become a pig in a poke. We needed the private money to build this thing. Besides, it's the water companies who have more knowledge about the world's water crisis than anyone. They understand what's at stake."

"I know. They have a lot at stake. Is Purity funding this?"

"They're the only ones funding it. It's in their best interests to be leading this new research."

"Bill, tell me what they'd gain by having free rainwater fall all over the earth. I don't believe you get it. The last thing they want is for anyone to create a way to seed ocean clouds. If we can coax ocean clouds to drop more body water over land, the water companies will be out of business. Think about it, Bill, why would they want free water for everyone? Why would they invest in this venture?"

"I get it, Julius. But without their money, the solution is impossible. Even their CEO—"

"I know their CEO, Steve Muller. He's a schlange."

"A what?"

"A snake."

"Call him what you want but his money is what's making oceanic cloud seeding a reality. Bernard Loren is his scientist. He would be your number two man."

"I know Loren. We've been on the same panel at water conferences. He'd never settle for second fiddle. Besides, he's a fake. An arrogant one, too."

"You may disagree with his science theories and papers, but he's considered to be brilliant in the field."

"Bill, I've known plenty of brilliant people. The ones that concern me are the ones without any common sense, or morals."

My ear was still close to the crack. I could hear Grandfather turn and walk away from this Senator. "I'll find the right combination of chemicals, I assure you."

"Of course, you will, Julius."

Whoever this senator is, he's very sarcastic. I snuck back upstairs.

I stretched out on the bed and thought about my parents. They'd know what to do. *Why aren't they here now? Why did God take them away? Do you even exist, God?*

– 2 –

I Come from Here

PEOPLE ASK ME ABOUT MY NAME, "Grayson." I'm named for my father, Gray Fields. You know, like "son of Gray." I suppose he wanted a son, but he and Mother got me, a girl.

I was an only child. "Gifted," they told me. But what would a freshman girl of fifteen at Georgetown really know about giftedness?

I'm now twenty-seven with a Ph.D. in math and science from Georgetown University; and proud to say I'm the third generation Ph.D. from Georgetown.

Like my parents and their parents, I loved chemistry. I spent my time reading and making new concoctions of weight loss pills which never seemed to work.

But after a couple years of Manduu fitness training, I'm now slim, fit, and determined to save the world.

This drought didn't just appear overnight. It came after a thousand unheeded warnings. I'd known about white water (*body water*!) problems since my childhood. My parents lived with the same life mission—to save the world. Until the world killed them in a helicopter crash.

They were both Ph.D. s with the National Oceanic and Atmospheric Administration (NOAA), specializing in cloud seeding. On their way to NOAA's cloud-seeding facility near Darjeeling, India, their helicopter crashed into the Himalayas.

Investigators concluded it was pilot error. I was young and confused.

My parents weren't normal. They were geniuses. I discovered the test results of my father's IQ in a desk drawer when I was much younger. 158. I had no idea what it meant at the time. I never saw my mother's. But hearing them banter back and forth over cloud-seeding methods, chemistry, algorithms, and the like, I was most certain she was on an intellectual par with him.

My father, Gray, grew up in Blowing Rock, North Carolina in the late 70s, the middle sibling of five.

He fell in love with chemistry while working in his father's pharmacy. After school, instead of playing any of the jock sports, he mixed chemicals. Not the illegal type but experiments with new compounds. His dream was to make new discoveries for more cancer cures.

In his junior year at Duke, he met Irene—also brilliant, also studying molecular chemistry. They married shortly after their senior year. Later, both received Ph.D.s in math and science from Georgetown. Cloud seeding became their passion, and together they worked in NOAA's D.C. office in the "Water Resource Sustainability" section.

I was born October 29, 2009. Pregnancy complications dictated I would be their only child.

I grew up in a moderate D.C. neighborhood, in a modest house, with progressive parents more interested in saving the world than making money. Irene and Gray Fields spent their time at home teaching me about the impending perils of the world's water resources. They taught me that science, and only science, had the answers to life's perplexing issues. Everything could and should be verified through scientific principles. "Ahh, they would say, the most exciting scientific principles are those yet to be discovered."

Gray Fields' work was his hobby. He loved what he did. Irene, too, but she had her garden club for a diversion. She was an award-winning horticulturist, but her true avocation was land and water conservation. Before the drought, she was the lead figure for wetlands lobbying on The Hill. She was able to obtain millions of dollars for America's wetlands.

Few people know that the Garden Club of America still carries such weight in Washington. It's only because their leaders are smart and savvy. My mother taught me more about our planet than I learned from books.

If cloud seeding seems unfamiliar to you, consider that it's not new science. It's been around for quite some time. It all began in the 1940s with America and Israel leading the way, soon followed by Africa and China. Cloud seeding is now used in 156 countries. Improvements to the process have been made over the years, but my grandfather discovered the most useful of all—his idea and experiments to introduce a special aeronautical mixture of buffered white phosphorus and liquid propane which created land-based rain clouds ten times faster and with more water droplets than the old methods.

Thanks to my grandfather, NOAA could take a decent cloud, inject certain chemicals into it, blow it up into a super vapor cloud, and drop much-needed water over land.

After my parents died in the crash, I moved in with my grandfather, Mother's widowed father. He's eighty and I assure you he's not only a genius, but a quirky one.

His patented invention increased production of food crops by 15 percent, worldwide. Impressive, given that for the last 85 years only incremental increases in cloud seeding science had been achieved.

But when the drought came, the tables turned. In less than a year our lakes, rivers, and aquifers—as far as the East is from the West—had evaporated and given up most of their life-giving substance. The misery began to affect everyone, gradually at first.

On the Great Lakes, and lakes around the globe, land squatters by the hundreds of thousands set up makeshift drilling rigs, digging through the mud cracked lake shores, searching for underground water.

But the *scientific* search for body water started with Grandfather's dogged attempts to discover the algorithm that would coax ocean rainstorms over land. There are plenty of rain clouds over oceans. The sun makes evaporation easy for the sea. But 90 percent of its clean and fresh water goes straight back to saltwater. Only 10 percent make it to land with their precious body water.

"Opa," as I affectionately call my grandfather, predicted the drought five years ago and so converted his land-based simulator into an oceanic cloud-seeding scale model. Seeing his converted simulator for the first time is a memory I will never forget. It was like walking into a college basketball gymnasium (which it was) and feeling as if you were inside a miniature eco-system (which you were) surrounded by glass on all four walls and a rounded ceiling of glass and metal supports. You had entered an environment that mimicked a portion of our earth's atmosphere. But the real show was on the court floor.

A platform, waist high, covered the entire gym. One-half consisted of hundreds of miniature cities, buildings, rivers, and farmland (with tiny grass, tiny row crop farms, dry ponds, and dry lakes).

The other half of the platform was water, representing the Atlantic Ocean with its little waves lapping onto the Eastern U.S. coastline from Maine to North Carolina.

I dipped my finger in the water and tasted. Yep, saltwater.

The doggone scene was so life-like I thought I should be seeing vehicles moving and people the size of ants walking down sidewalks. Instead, it was all so static; still, it was a laboratory instrument larger than any I'd ever seen.

It was so elaborate I swore I was looking at an actual slice of the east coast. New York City had its five boroughs, along with the Hudson and East Rivers. The miniature landmarks—the connecting borough bridges, the familiar skyscrapers, even the gift-shop-purchased Statue of Liberty—were all there. The entire spectacle looked like a giant train set, without the trains. There was even a tiny Liberty Bell in Philadelphia.

My grandfather's international patents had brought wealth to him and helped replenish the coffers of Georgetown University. But Opa was never interested in money, only science.

As one Op-Ed article in a 2034 New York Times asked, "What's the only substance on earth essential to life that can be bought and sold? Is it gold? Plutonium? Food? Not even *food* can be produced without water. Water is the one essential element we have abandoned."

The consequences have been manifold. Ugly territorial wars break out every day in what are known as hot stains. War lords from undeveloped nations are especially evil—raiding and pillaging villages for their water; what little there is.

In developed countries, the water rights war began in the court system. Many rivers and aquifers flowed across borders.

Legally, whose water is it? Half claimed the "up-stream", or "up-river" water belonged to those downstream as well.

But those wars ended. Court battles became moot when there was no water to fight over.

Enter a new era of opportunistic people. Water supply fell into the hands of water filtraters and distributors who first controlled the supply of water through their in-home grey water filtration systems. Grey water is still being recycled inside homes for sinks and showers and washing machines, just not safe for drinking.

Purity Worldwide became the largest manufacturer of in-home grey water filtration systems. It's how they grew to $120 billion.

Soon came the bigger prize. Water filtraters bid for and built micro-plants in municipalities around the globe, each using the chemicals and membranes that can turn black water into clean body water. Since 2005, this black water filtration process has been able to produce clear potable water. No bacteria (supposedly). No impurities (supposedly).

My Ph.D. dissertation was on black water filtration. I was twenty and finishing my doctorate when I began to understand why Opa started working on oceanic cloud seeding instead of land-based seeding.

My entire family understood that black water filtration was going to have problems, sometime in the future, all because demand would eventually outstrip supply.

At that time, no one else was working on building rain clouds close to shore and getting them to travel over land to supply fresh body water.

Grandfather's first simulator experiments began back in 2015 when the university granted him their old Hoya gymnasium to experiment with land-based cloud seeding.

He made his greatest discovery there—his white phosphorous method for super heating surface water (lakes and ground water) so clouds could form faster. His chemical algorithm was put into real world use and was an instant success.

But after two years of the drought, there was no need for land-based seeding. Without lakes and rivers and ground water there was no evaporation. Without evaporation there were no clouds.

Grandfather soon realized that his land-based surface water phosphorus could be put to use over ocean water. It was a theory. And certainly doable; but for it to work, he had to figure out how to manipulate an offshore cloud to come to earth.

The science paper that won him the grant to retrofit his simulator for oceanic cloud seeding was titled: 'Coaxing Ocean Clouds Over Land.'

Grandfather was convinced he could increase the amount of ocean water dropped over land by 15 percent. That would indeed be more than enough to win the war on water depletion.

So, you can see why this is my family's story as much as it is mine.

– 3 –

OPA

THERE WAS A SOFT KNOCK AT the door.

"Yes?"

"Madam, it's Arnold. I'm here to draw your bath. It's bath night."

"I know. Don't bother; I'll do it. But thank you, Arnold." I had said that to him a million times.

Ten minutes later I slipped into a half-full tub of lukewarm recycled water.

Afterwards, I dressed and stretched out on the bed. I remembered how Grandfather had spoken with disdain about Steve Muller.

I opened the bedside drawer and pulled out a book, "The Impending Water Opportunities" by Steven Muller. My bookmark was a piece of torn newspaper from the Sunday, June 12, 2033 Washington Democrat. It was a small side bar to the major story.

PURITY WANTS MORE POWER

In a related story, the House Committee on Water Resources is considering yet another bill that would give more authority to America's largest water distribution giants, *Purity Worldwide* and *American h2o*.

Purity remains the largest water distributor in the world with over $400 billion in sales. *JALA India* and *Beijing Water* remain state owned distributors.

Purity's CEO, Steve Muller, claims that his company needs new pricing authority in order to invest in R&D to supply 'body' water to more underdeveloped countries.

Most Democrats wanted to nationalize the water industry—take it out of the hands of private companies.

Senators Bennett, Weinhold, and a few others pushed the hardest against nationalization. They were considered outcasts in their own party, purely because they were siding with the Republicans. Many Democrats, along with Republicans, suspected Bennett and Weinhold of being in the back pocket of Purity Worldwide.

We all knew the water companies would do anything to avoid nationalization. It would put them under the thumb of bureaucrats.

So, one of the bills could nationalize water resources and neuter the water companies. The other bill could give them even more control over pricing if they stayed private.

These were two divergent, opposite sides of the political debate. Both bills were in congressional committees. And these bills were the talk of the town, the D.C. gossip merry-go-round.

It made me wonder about Steve Muller's strategy. Surely, he kept the congressional cookie jar replenished with plenty of bribe money. Somewhere in the past I read where he had learned this strategy from his father, "Big Tom" Muller, late founder of Purity.

At seven sharp my watch sounded a soft tone which meant dinner time with Grandfather. I always enjoyed talking shop with Opa. My five years with him had taught me much.

And, I had long since stopped asking him about the circumstances of my parents' deaths. But somehow, I always felt he knew something about the how and the why. He just never was willing to tell me.

Even though I'd been living in what I considered luxury for the last four years I never put any stock in the glory of riches, just the convenience of it. Maybe that's because I lived with so little growing up. Gray and Irene Fields were salaried scientists at NOAA.

Grandfather became famous after he invented, patented, and made a fortune from the right combination of gases and minerals that cut in half the time it takes to seed a cloud and make rainwater. Cloud seeding took over like never before. Large portions of the earth's deserts became fertile grounds again.

But when the drought began, land-based cloud seeding quickly became a lost cause. Even the lack of snow melt meant no more evaporation to create large enough clouds for seeding.

Chemicals existed to create larger and larger ocean storms, but there was no method for coaxing them over land.

"Ahh, science," he would say, "there-in lies the truth, and the quest, my dear. Unfortunately, it takes money."

When other scientists, students, authors, and the halls of congress visited him at 5804 Highland, they invariably considered him rich. But that wasn't the reality. All he had was the mansion, the five-car garage for one car, Jesse the bot chauffer, and an annuity to live on. That was about it. Most of his fortune had been spent on the simulator.

After his notoriety as a famous Georgetown professor became truly worldwide, a wealthy alumna, steeped in environmental issues, bequeathed her historical home to him. Opa never would tell me if they had some sort of relationship.

Unfortunately, after she passed, none of Grandfather's money went into repairing and maintaining the place. It was drafty, damp, and dreary.

And he was eccentric. He still looked and dressed in a dreary sweater under a rumpled tweet jacket, like a slovenly rumpled professor. Even when he spoke at water conferences around the globe as a celebrated scientist, he never bothered with appearances.

He was quite simply the epitome of an eighty-year-old German giant with stooped shoulders, droopy jowls entrenched behind a thick white beard with plump, pink lips—due, no doubt, to an ever-present and hearty smile. He has always reminded me of some aged movie star, though I could never figure out which one.

"Grandfather, the oysters were delicious. And the fish was exquisite."

"Well, we can thank your mother's saltwater reclamation project for that."

My mother, Dr. Irene Schwarzkopf Fields, had led the Garden Club of America's Chesapeake clean-up fundraiser that saved the bay along with the oyster population.

But tonight, mother wasn't on my mind; someone else was. "I heard you and the Senator talking this afternoon."

"I would expect no less of you than to spy on me, my dear. What did you learn?" He sucked an oyster from its shell.

"I wasn't spying, Opa, merely eavesdropping."

"Oh, I see. So, what did you learn?"

"Not much. But I think the Senator is a wimp."

"A wimp?" Grandfather's laugh was always hearty and contagious. "My dear, you are absolutely on target."

My coy smile and snide question brought him to attention. "So, what's his name?" I asked.

Opa cast a wide smile, lowered his fork of fish, and with a twinkle in his eyes asked, "Why would you wish to know, Grayson? Would you try to meet with him and let him know he has no backbone?"

"Ha. No, just curious." Of course, I was fibbing.

"If you must know, my child, his name is Bill Bennett, Senator from New Jersey. Now, let's change the topic. I want to talk about the simulator test. Will you be there to help next week?" He picked up his napkin and continued to eat.

I cocked an eyebrow and grinned. "Of course, you know how much I enjoy trying to make it rain over the Atlantic. How many times as it been now? Two dozen?"

We both paused to eat more fish. "Who's counting?" He asked. "Besides, my dear Grayson, you realize how close we are, don't you? The algorithm is so close I can almost taste it." He held his fork of fish up and savored the bite.

"Are we? Are we close?"

Whenever Grandfather was disappointed in me, he'd push his lips up into a smirk allowing his big eyebrows to furrow in the middle. "Don't be sarcastic with me, my dear. You know I have no patience for naysayers."

Time to change his mood. "Opa, have you ever met Steve Muller? I read his book. He seems like a decent guy."

"His book is all bluster, and he has an inordinate fascination with himself. He's anything but a decent person." Opa's eyebrows squeezed into a stern look. "He's a snake with fangs that encompass our entire world. I consider him to be Satan's minion."

"Goodness, Grandfather, I was hoping to put you in a good mood."

"I hope you never know him. For your sake."

Obviously, Muller's view of himself in his book, 'The Impending Water Opportunities" had not been as factual as advertised.

– 4 –

STEVE MULLER

EIGHT COMPUTER MONITORS OCCUPIED Steve Muller's enormous u-shaped aquarium desk, fifteen feet across the front and sides.

Dozens of beautiful fish swam inside the transformed aquarium/desk. The soft sound of percolating bubbles was relaxing.

But seeing Muller sitting like he was half in and half out of an aquarium was not at all comforting. His legs seemed to be swimming under water, among the fish, while the blue aquarium lights shown on his fifty-year-old, surreal handsome face.

Steve Muller believed a lavish lifestyle was one way of intimidating subordinates and adversaries into believing he had the power to move mountains. And he did. His eyes were set on the eastern Himalayas. After all, he was the wealthiest man in the world—not just according to him, but every financial news outlet that kept up with rubbish like that.

As predicted, water became more valuable than oil, diamonds, and gold combined. Muller's nickname in the media was "Poseidon," as in the Greek's Olympian *god of the sea and king of the sea gods.* He was proud of it. When he met with any new subordinate, a competitor, even a Senator for the first time, he would push a button under his desk, sending bubbles and new fish swimming eagerly up through a clear tube into his desk/aquarium.

As the new species of fish entered, he'd casually mention, "Did you know piranha are the most efficient meat eaters in the world? Even sharks can't compare to the savagery of piranha. Piranha can clean the flesh completely off the bones of their meal in seconds."

Steve Muller enjoyed watching the reactions of his guests as they stared in horror whilst the desk transitioned from aqua blue to dark pink, and tiny fish skeletons drifted to the bottom.

For Muller, everything was about power and intimidation.

By morning, the tank was cleaned and replenished with new, exotic freshwater fish.

"Steve, I tried," said Senator Bennett, begging. "But Schwarzkopf's not budging. Says he can discover the solution at his—"

"Bill, Bill, you're such a disaster. How did you ever get elected? I suppose I'll have to do this myself. I'll get him over there."

"But Steve, I don't believe it matters. He'll never figure it out with his antiquated simulator. He's out of money. He'll fail on his own turf."

"Now Bill, what assurances can you give me that he'll fail? Huh? None. No assurance. I've told you before, we need him to fail on our turf, not his. NOAA needs to witness it, so congress knows. Remember? That's the plan. Unless we have him under our roof, under our watch, we'll never know in time before the nationalization bill comes up for vote, will we? You disappoint me, Bill. Just wait for further instructions until I figure this out."

On Muller's left, a large wall monitor came to life with a message.

`HOT STAIN REPORTS IN ONE MINUTE`

"Bill, I have to go."

"I'll keep working on Schwarzkopf," Bennett said softly. "I'll get him to the NOAA facility, Steve. I promise."

"No, back off. I'll figure it out myself. And, Bill, it's no longer called the NOAA facility. It's now the *Purity*/NOAA facility. That should show you how important this is to me. You understand, don't you?"

"Yes, I understand." Bennett sensed his body beginning to sweat.

Muller hung up and leaned back in his executive chair, waiting for the screen's next message regarding hot stains.

Hot stains came into existence in 2002. Each represents one of the most depleted water spots on the planet. One of

Muller's advantages was that he had operations and spies on the ground in most hot stain locations.

Other water companies, like American h2O, Avanti, MexCo, even the World Bank, the Federation of Nations, and WHO knew about these hot stains. But Steve Muller had his own *private* updates—far better intel than any others could imagine. For Steve Muller, any one of these hot stains could mean billions more in revenue for Purity.

Every week, he watched his spies from around the globe deliver updates. They were high-ranking officials in their respective countries. Their government salaries were a mere pittance compared to the exorbitant money Purity paid.

The reports were delivered via live feed on a giant wall monitor to his left; each was a condensed version of current activity. Muller liked them short, to the point, only highlighting the most pertinent facts.

The name of the city—or country—appeared on the big screen as each report came in. They all began with a live image of the employee's face, alongside videos, and photographs taken at the location.

GO AHEAD FOR LIVE REPORTS

SCOTTSDALE appeared on the screen above a man's face. "Mr. Muller, we're going to cover only six hot stains this week. First up is Scottsdale. I'll give that report myself, sir. Are you ready to begin?"

Roberto Orci was Muller's chief foreign officer. He managed the entire operation of hot stain 'spies.'

"Go ahead, Roberto."

"Just yesterday, the angry mob you see on your screen in Scottsdale overwhelmed several wealthy neighborhoods. Reports indicate about four areas were targeted. It appears the mobs are poor immigrants complaining about inferior bulk water delivered to water stations. The mob broke into some homes, stole water containers, and are still holding some of the homeowners hostage. The police seem to be getting it under control. Our plant was not affected. Any questions, sir?"

"Have you met with the mayor?"

"I have. He's wanting an answer from us about the water supply. He's already complained several times to Senator Gill."

"Tell him we'll send six more containers next week. Call him. Tell him now."

Muller pushed a key on his computer.

A new face and city name appeared on the screen: MEXICO CITY. "Imelda Sanchez here, sir. Just to give you some background, we recently reached 30 million people population. The smog is killing five thousand a week, though. It's pretty awful. Since the drought began, we're losing over 250,000 a year. Starvation and disease, sir. I believe the outskirts of Mexico City may be one of our worst hot stains. The government is beginning a campaign to blame U.S. water companies for exporting inferior water. Our market share here is only 12 percent, sir, so we're predicting Purity won't be the target. It'll be American h2O. While MexCo has the largest market share, President Rodriguez will never blame his own company. He'll blame

American h2O. The sanitation systems are deplorable. It's not a pretty picture, sir. I'm meeting next week with the Minister of Interior and hope to secure a contract for Purity to replace American h2O. I'll make sure that happens. That's all for now, sir. Any questions?"

"No questions. Just get the best price you can. And Imelda, promise nothing but pure water."

NIGERIA was next. "Mr. Muller, I'm afraid we have not-so-good news report from Nigeria. We're still in a dramatic state of decline. Earlier this year, I reported our population had peaked three years ago at 20 million. That number is now 18 million, due to continued starvation and tribal wars over water rights. Sir, I'm requesting that we pull the Purity team out. I know this is drastic, but it's getting dangerous. I'll submit a full report on this in a few days for your review."

"No. Get them out now. There's no need for another report. And Ray, abandon the equipment and plant. It's okay. Insurance will cover it. Report back to Roberto. He'll have another assignment for you."

BUENOS AIRES. "Buenos días, Señor Muller. The Paraná River, the lifeline for Buenos Aires, is approaching complete dryness now. The city and the country are in chaos. Our Purity tankers are being hi-jacked at the port. (While the man is talking, video is revealing scenes of his report.) The thieves are shooting anyone who resists them. Some government buildings, just last week, have been overtaken. A full revolt is underway. The hijackers are from the countryside. They're taking the victim's IDs to find their homes so they can scavenge more water. We've hired more security at the port, sir, and I can assure you we will regain control in a matter of weeks. Señor Muller, I listened in on the Nigeria report a moment ago. I am not recommending abandonment here in Buenos Aires. Not yet. We *will* regain order, sir."

"Good, Paule. Keep me updated if it gets worse."

DELHI. An Indian woman appeared on the screen. "Mister Muller, am I on?"

"Go ahead, Nipa."

"Reporting from Delhi. Not much has changed since my last report two weeks ago, sir, except that in the north, the snowmelt has nearly vanished now. The rivers flowing from the Himalayas are continuing to grow weaker. I'm afraid the people in the north are now forming tribes. That's a new development. I've received reports that war lords are taking over some of the provinces. It will become ugly there. I'm afraid it's much worse here in Delhi. The drought continues to deplete the aquifers, and hundreds of wells a day are pumping up mud. I'm recommending that we begin shipment of additional tankers to all the southern ports. Will Purity be able to supply two and a half million gallons a week?"

"We'll get on it. Have you negotiated a price with the government?"

"Not yet, sir. I'm working on it."

Very good, Nipa. Keep me informed."

Similar reports came in from Los Angeles, Dallas, Cairo, Baghdad, Kampala, Paris, and London, all cities where Purity has purification plants and can make more money than many countries.

The screen went black for a second and a new message appeared:

(*Six Reports Unavailable*)

THIS CONCLUDES THE WEEK'S BRIEFING.

Muller wasn't surprised by the *unavailables.* Those "spies" were often busy with more pressing matters.

Many were dealing with land-locked countries where getting clean body water had become a fight for survival. Without a port to receive treated water, they were the most vulnerable. Even countries that once had life-sustaining rivers had lost their lifeline.

Births continued to exceed deaths; our planet had ballooned to over nine billion people. Yet, scientists predicted our earth could sustain only nice billion.

– 5 –

KUPUP, INDIA

IN 2031, LONDON'S ALL SAINTS CHURCH sent a new missionary and his family to Kupup, India. His name was Brooks Turnage, an African-descended Brit who had distinguished himself in the British Infantry.

At age thirty-eight, Turnage became an ordained minister and set out to fulfil his next calling—the mission field. He came with a wife, Anna, two daughters, and a heart full of hope. Brooks and his family had been in the small village of Kupup for five years. Their home was a modest three-room concrete block building with no running water or decent amenities. The village couldn't afford grey water recycle appliances. Worst of all, their water well was diminishing by the week.

Only 75 percent of India was fairly-well educated, while remote villages like Kupup were still living a nineteenth century agricultural existence. Located in extreme northeastern India, Kupup was surrounded by Bhutan a few miles west, Bangladesh to the south, and the "giant ones"—the Himalayas—a short distance to the north. The snow caps that once flowed down from the giant ones were no more.

One Saturday night Brooks and Anna were in bed, tired, sleep not far off—their two daughters, Lynn and Lisa, already asleep in another room.

Brooks whispered his frustration to Anna, "Tomorrow will be just like any other Sunday. Only the same ones who have nothing else to do will show up. Five years, and I'm getting nowhere. Not a single convert," he sighed.

"You're forgetting Rinku. And I'm sure there are others you just don't know about."

"Rinku doesn't count. He's like family. He's so dependent on us he would do anything we asked."

"I don't agree. I've witnessed his spirit grow. He's learned more than you realize. Now get some sleep," Anna said, turning under the covers. "You have a sermon to preach tomorrow."

Rinku Patel was an eighteen-year-old orphan, who had taken up with the Turnages after his parents died five years ago from the tainted water disease.

Sunday, noon. Kupup's Circle Leader never allowed Brooks to hold church before noon. The mornings were for the villagers to work the meager crops, milk goats, tend to the water well, and do as much as possible before the sun's oven could reach baking point.

The church tent was old and faded. A pole in the middle kept the canvas propped up; poles at each corner with tie-down ropes completed the church—except for the driftwood Christian cross that hung behind Brooks' podium. Pews were made of tree stumps, long wooden planks stretched between them. Missing were any walls to the church. The activity outside the tent was always a distraction for Brooks. And today would test his powers of concentration.

A portable music box played recorded hymns, signifying to the villagers that church services were beginning.

Brooks, Anna, the girls, Rinku, and eight elderly villagers sang hymns. When the music stopped, Brooks stepped behind his podium.

"Brothers and sisters, in the name of the Lord God Almighty, let us pray: 'Father of all, creator of the universe and the only one true God, we give thanks, first for the water you've provided us. We also give thanks for these men and women who thirst for you. We are knocking at your door now, asking that you keep our village safe, asking that you would bring souls to you today. Yes, Lord, today! I don't have the power to save souls, only you do. So, I ask that Holy Spirit would put the words in my mouth today which would glorify your Holy Name. Amen."

"Amen," from the congregation of twelve, most of them infirmed, not able to contribute help with any village chores.

"As I prepared for today's sermon," Brooks said, "I was reminded of Nehemiah, whom God chose to rebuild the walls of Jerusalem. Nehemiah was first and foremost a man of prayer. He prayed for four months for success to build the wall. That was before he even lifted a finger to start! Just as we here in Kupup need a new water well, we need to pray like Nehemiah. We need—"

Brooks' attention was diverted to unusually loud noises outside the tent, which quickly turned into a village-wide cacophony of villagers making their way past the tent.

Brooks tried to continue, "We ask you Lord God, to provide for our needs, as David prayed for your help, we also pray."

Brooks, nor his congregation could keep their eyes or minds off the activity outside the tent. "All of you, please sing this next hymn while I find out what this is about."

Rinku went with Brooks. Anna and the girls watched them stop to talk with Ajay, one of the elders.

Anna lost sight of them as the villagers continued to head for the hill—the hill where Manu, the Circle Leader, always gives his pronouncements concerning village matters.

"Ajay, what's going on?" Brooks asked the old wise man.

"Rashman has returned."

"Good. He's been in Anani for a while, hasn't he?"

"No, Reverend, he's been in Bhutan."

"I was told he went to Anani to visit a sick cousin."

"It just goes to prove that the words of men are not always from the truth."

They walked on along with the crowd.

Brooks noted villagers hustling past him to secure a good spot on the hill. "So, tell me, Ajay, why is Rashman's return causing such a stir?"

"Sturr? Ahh, you British have such funny words," Ajay said. "What does sturr mean?"

"It's like gomlamAla. A commotion," said Brooks.

"Very well; now I know sturr."

They continued moving with the crowd.

"So, what's it about?" asked Brooks. "Why is our village so excited?"

"The sturr is not good news, I'm afraid."

Brooks turned to look Ajay in the eye. "Then there must be bad news."

"Ahh, be patient, Reverend Turnage, Manu will tell us soon enough."

As they and hundreds of villagers reached the natural amphitheater many were already seated. Manu was at the bottom of the hill, standing on a wooden platform, stamping his six-foot-long staff to get the crowd's attention. *Rap, rap, rap.* The sound reverberated up the hill. The crowd quieted.

"Rashman brings us disturbing news," announced Manu. His deep voice rolled up the hillside for all to hear. "The Bhutanese are coming. Coming here to take our water." The crowd erupted in a loud murmur. He rapped his staff three times.

Again, there was complete silence. Jaiman, Manu's son, stood near the platform. He spun around, bowed his back, and held a defiant fist in the air towards those on the hill. "We will defend what is ours," he screamed. "These

Bhutanese mongrels wish to take our water. Our lives! They will not. We will fight! And we will beat them back."

The crowd cheered.

Jaiman was a rare individual, one with enough charisma to turn his own fervor into the crowd's will.

Manu spoke again: "Circle members! We will meet in fifteen minutes in the main hall." This meant plans for survival were going to be discussed and assignments made by Manu.

The crowd erupted with cheers of fighting spirit. They broke into small groups and began speculating about the night ahead.

The circle members consisted mostly of Manu's relatives—his sons, brothers, an uncle, nephews, and four others who exhibited wisdom—according to Manu.

Jaiman took advantage of the break to search for his fiancé, Anashka. When they met, she was distraught.

"Jaiman, our wedding!"

Holding her face in his hands and looking squarely into her eyes, he said, "Nothing will come between us and our wedding. I promise you. But now, I must attend this council meeting." He kissed her forehead, hard.

She reached up and kissed him on the lips even harder.

They parted.

Rashman caught up with him on the path, worried about something. Something other than the defense of the village.

"Jaiman, stop. Listen to me."

"You know I always listen to you. What is it?"

Rashman grabbed Jaiman's arm and drew him off to the side. "Listen to me. After the ceremony I want you to take Anashka and leave immediately. Go to Anani for one week. You know my sister there, Binita. Wait for my word. If you don't hear from me, do not return here. Do you understand me, my friend? Do not return."

"I hear you, but I don't understand. What are you *not* telling me? Because I am not going to disappoint my father! Anashka and I will have the traditional ceremony. The mandap. All of it. It would break his heart if we didn't."

"Listen to me. Like you, I too am in love. And I know—"

"You're in love? You didn't tell me! That is the best news." Jaiman cast a broad smile and asked, "Who is she?"

"You don't know her. Just listen to me."

"No, I won't. And what do you mean I do not know her? I know everyone in our village."

They stared at each other for seconds, each wondering what next to say.

Brooks, his girls, and Rinku walked up.

Brooks was first to speak. His harsh, military tone rattled Rashman and Jaiman. "Rashman, where is this village you've been to?"

Rashman responded. "Why, Reverend Turnage? What would you intend to do? Go there and fight them?"

"I believe I can stop them from coming here, yes."

"Stop them? Are you what your British call 'a bloody fool?' Listen, I spent over two months there. It's a village only Manu knows of. Do you remember the dying man who came to our village three months ago? Do you remember him?"

"I only remember a man coming from another village and dying here."

"He was from Gedu in Bhutan," Rashman said, pointing over the heads of Anna and Lisa, "there, not many kilometers. Maybe two days. You don't know the Bhutanese. They can be savages. The dying man told us of their water problem. And their plans. He had been expelled from their village because of his age.

Manu sent me to look on them. To find if the man was right. He was right. We are one of the villages they talked of. I became the spy for our village. They know of our well."

Brooks turned to his daughters, "Lisa, Lynn, it's time for you to go home to your mother. Do you understand?"

"Yes, father," Lynn said. They dutifully turned and walked away, glancing back along the way, concern on their faces.

Jaiman spoke this time: "Reverend Turnage, if you truly desire to help us, you can join the men on the boundary who are digging trenches and preparing the pits."

Rashman pointed at Brooks and asked, "Tell me, Reverend Turnage, have you ever charmed a Cobra?"

"No. But listen to me, please." Brooks was begging.

Rashman wasn't listening. "You see, it is not easy to charm a cobra. You must first become one with the animal. Oh, but the cobra, you must show him you are a stronger animal than him. It's the lure of the music. The flute. For the Cobra it's what you British call the 'pied piper.' He can't resist. He will be docile and obey you. But if he doesn't, you still have the advantage. You have drawn him into your trap. You can cut off his head. And that is what we will do to these Bhutanese mongrels who come for our water."

"Bloodshed is not necessary. Give me directions to the village. I believe I can show them how to dig for a new well. As I did here. Dammit! Give me a chance."

Laughing, Rashman waved an arm between them, as if swatting a fly. "There is nothing you can do, Reverend. But we admire your bravery, and your passion for cursing. This would be a good time for you to take your family back to England; if you care for their safety."

Brooks stood dumbfounded. Jaiman and Rashman turned and walked with others to the meeting hall. Brooks watched them for a while, pondered, then slowly lowered his head in defeat. He put his arm around Rinku.

"Rinku, we are not invited to the council meeting. There is nothing for us to do. Except pray."

Brooks and his family helped the villagers prepare for a likely attack. Dozens of men and women dug trenches around the perimeter. It was a three-mile circle, so they did their best in a couple of days.

In the dead forest above the village, several deep holes were dug and covered with branches to disguise the pits. Cobra handlers placed two or three snakes from their cobra farm in each pit.

Closer to the village, booby traps of leftover dynamite were placed. Jaiman and some men pored oil and gasoline into trenches. Torches were assembled using a stick wired to a can. The can was stuffed with animal skins saturated in oil. When lit, they would be used to light the trenches at first sign of the enemy.

Sticks, hoes, shovels, any long object, were sharpened to make spears. They had no modern warfare. Jaiman organized 24-hour guard duties, not knowing what day, *or even if*, the Bhutanese would attack.

They had no idea what would befall them.

– 6 –

THE SIMULATOR TEST

JESSE DROPPED GRANDFATHER AND ME off on the Georgetown campus. We walked a few hundred yards across a wilted Healy Lawn to a fortress-looking circular brick building attached to the old Hoya basketball arena. This newer, attached structure was meant for extra security.

Grandfather punched some numbers into a keypad, then placed his eye against a recessed screen built into the brick wall. A non-descript, thick metal door slid sideways into the wall. We walked in and Grandfather said "Hello, Joseph," to a private security guard who'd been there since I could remember. He led us to another steel door, where he and Grandfather both placed their hands on a console to open another door. The guard left us as we walked into the converted gymnasium—the gym the school had awarded to Grandfather in 2015 for his original land-based cloud seeding experiments.

Years ago, when I first entered his secret laboratory, I was mesmerized by the sheer size of the cloud-seeding simulator. It's still impressive. Half of the waist-high platform is covered with saltwater representing the Atlantic Ocean. The other half, part of the eastern U.S.

But the old gym bleachers have been replaced with canisters of chemicals, tables for computers, and offices for Opa and his grad students.

I noticed new computers, new high-tech fans, and larger mixing containers all along the sidelines, only a few feet from the platform.

"Opa, when did you install these upgrades? I'm impressed."

"When you weren't here helping me, my dear." He smiled.

"I'm sorry." I felt bad that I hadn't been here for several months. Curious, I asked, "Where did you get the money for all this?"

"I used the last of it. All I have. All we have. I'm sorry. You may not have any inheritance, my dear."

"No, don't be sorry. I don't care. You know I was taught not to care about money. C'mon, let's see if we can make it rain with this beauty of yours."

"Yes, indeed, I suppose she *is* my 'Grand Damsel'," he said in his cavernous German accent with a laugh and a smile that stretched across his cheeks.

He continued, "She can create big rain clouds over my miniature ocean. I've done that. The larger question now is (he pointed at the Statue of Liberty) can we coax her to move over that! Can we bring her to land and then make her give up her precious pure water? That my dear, has been the conundrum."

"Hmm," I muttered, having only a perception of his theory from what he'd told me before.

"My dear, I truly believe I have discovered the ultimate connection between certain ions. Not just the gravitational forces between positive and negative, but the use of ions in the atmosphere to influence their behavior; like when stronger positive ions being sprayed over land connect with negative ions over water, they want to bind together like two lovers. Surely you remember the principles of magnetism."

I could only scratch my head and smile with wonderment at his genius and fervor.

He continued, "But I must determine the correct formula and flow rate of the positive and negative ions to succeed. Today we're going to experiment with a new algorithm I tested on the computer."

He sat down at the computer and powered up the system. He had the excited look of a six-year-old boy about to launch his first rocket into space. (It's a "high" I'm convinced most scientists experience just before giving the go-ahead for a large-scale lab experiment.)

Everything must happen in perfect sequence. First, the sun lamp had to heat our miniature Atlantic Ocean. That takes four to five hours, just like here on earth. Late afternoon clouds tend to form. Not all are ideal for seeding.

Nozzles along the ceiling (representing airplanes) would spray the proper mixture of silver iodide, sodium chloride, and propane into the most promising cloud.

To speed up the process, Grandfather had discovered and patented the use of white phosphorous added to the chemical mix. It sped up evaporation, making it possible to increase a cloud's intensity by ten-fold over old methods.

Once one of those ideal clouds gained some heft, Opa gave me a thumbs up and yelled, "She's ready. There are two large stainless-steel canisters at the far end, over there," he said, pointing to the opposite end of the simulator. "I need you to station yourself between them. When I give you the word, turn the container valve on full force, count to ten, then turn the other container valve on. Is that clear?"

"Got it." Trotting around the miniature world, I almost stumbled with excitement. I always get excited when I'm in the simulator area helping him with an experiment. Part of my adrenaline surely came from realizing that we were inside an enclosed environment of bombs—the propane tanks, the potassium chloride, pressurized containers underneath the platform and along the perimeter. I had to block that out of my mind. Still, I knew to expect lightning strikes at some point.

"My dear, let's see if we can make it rain over New York City," he said.

"Are you using the same formula?" I asked, running around the platform.

He shouted, "I had to make a few changes. The important thing will be the fan wind rate for the propane. You know what to do with that, right?"

"I think so."

"Opa!" I shouted. I continued running. "You make everything seem so simple."

"Ahh, but once understood, science is like that."

As I ran, I saw him move over to a large box, open the lid, and scoop out several cups of what I knew were ammonium nitrate prills. He was about to fertilize the cloud. He poured this into another container.

I reached my designated station and could see him across the cavernous gym as he popped the lid off the powdered urea and added it to the mixer. I'd seen all these procedures several times before in our many experiments with land-based cloud seeding. Next, he poured a beaker of red liquid iodide into a smaller stainless-steel cylinder, capped it off, and pressurized it.

"Ready to run the program?" he shouted.

I was nervous. I could only nod. My computer was on a table behind me. I turned and pushed one key on the laptop.

The ceiling nozzles began to spray its reddish silver iodide down into the cloud. More nozzles released the ammonium nitrate.

"Now!" Opa shouted at me.

I opened the first valve, counted to ten and then opened the other one. From the nozzles located at base level (waist high) came a white, dry-powdery spray—non-toxic, water soluble phosphorous over the water. The use of phosphorous was the twelve-year-old discovery of Opa. It's designed to rapidly heat the water's surface, making evaporation occur faster.

I left my station and ran to him. Soon after, the hot phosphorous hit the water. We both began sweating like Bikram Yoga fanatics. The rising evaporative mist added to the cloud. It kept growing.

After thirty minutes the cloud nearly exploded. It began crackling with lightning strikes in its upper atmosphere.

Startled, we moved back a step. We watched two dozen nozzles around the top perimeter spew out a hissing cloud of vapor into the cloud. Larger lightning bolts scared the hell out of me. The cloud continued to grow. Bigger . . . Bigger . . . Bigger. The professor turned the last valve—the valve that opened hundreds of tiny needles that emitted a smoky white substance that made our Atlantic Ocean and the land look like some mystical fog had come from below and now moved like thick smoke swirling one inch above the entire platform. Opa looked up at the cloud and said, "Grayson, look, she's becoming our little giant, isn't she?"

"Oh yea," was all my lips could say. The cloud was thirty feet above us, dark and daunting. I thought twice about running, getting both of us out of the room.

"Now for the secret ingredient, the magnetic ions," he announced.

When Opa punched a key on his computer, a dozen pipes, each less than an inch in diameter, emitted a strong mist from both ends of the platform.

"There on that end," Opa said, pointing, "those are the positive ions being sprayed over the land. On the other end are the negative ions. Hopefully, in a few minutes the positives will meet the negatives and bring them to land. I only wish we had drones to do this work."

"One day you will, Opa. One day."

We waited in anticipation.

"Look, she's beginning to move in," I shouted.

I watched as the Manhattan-bound cloud continued inward. I muttered quietly, not wanting to jinx it, "Keeeeep going … Keeeep going."

Soon, giant lightning bolts crackled the water and shook the platform. We both backed up for fear it could strike us next.

"Should we step outside, Opa?"

"No, dear, just stand back. Look at her move."

Suddenly, before reaching land, the cloud crackled again, then unloaded. It was a miniature torrent of rain over a miniature Atlantic Ocean. The saltwater mist enveloped us, along with our devastation. We shut all systems down. We sat, heads shaking side to side, wondering why.

"Why? Why did this happen, Opa? I'm sorry," I said, "maybe I miscalculated the fan speed. Should we try it again?"

"Not today, my dear Enkelin. I need to adjust the algorithm somehow. I'll have to study the results. It's possible we should seed the cloud after it moves over land.

I'm not sure. Plus, many of the needles for the phosphorus are old. My biggest concern is with the vacuum seal. I'm not sure I'll ever get a good seal in here. She's just old. Old parts. Maybe we try again next week, huh?"

"Give me an idea of what a new simulator would cost? From scratch. What would it take?"

"Oh, Grayson, there's no way. I couldn't possibly find the kind of money we'd need to upgrade it."

"Just tell me!" I was upset. I could see failure in his eyes, and it hurt.

"Come, let's go to my office," he said, smiling gently at me, as if I were the only person in the world. I loved him for that.

I watched him bring up a spreadsheet on his computer, a line-by-line item of everything a new simulator would need. He had already attached a cost to each item. He printed it, looked at it, then handed it to me.

I took a look and dropped it on the floor. We both stared at it.

"Opa, we'll go to the university. They have a huge endowment."

"I've already asked. They said no." He gently reached over and took both my shoulders in his huge hands. A tear rolled down his cheek. "They said NOAA was going to solve oceanic cloud seeding in that damn facility in the Himalayas. And I spent your inheritance." (I'd never seen him cry.) "I'm so sorry, my dear."

His head sank in defeat. I embraced him. Heartbroken for him.

After a minute, I broke away, and looked into his eyes, only inches away, "I've applied for a job at Purity."

His eyes smiled tenderly, like he'd heard a little girl's prayer for something he knew wasn't possible. "Oh, no my little one, no. You cannot do that. You must stay away from him. Do you hear me? He's too dangerous."

"Opa, I can handle myself. I'll take all your notes and these videos. I know enough to go there and make it work if you stay here and figure out the algorithm. You can get it to me. You don't even have to go over there!"

Like him, I had tears in my eyes. "I will. I will get over there and with your formula, we'll figure it out. Trust me, Opa." My determination felt like a "life's calling" from a source I'd never met.

He pushed me back and grabbed my shoulders again.

"My little Grayson, I cannot let you do that. I've been withholding something from you in case of this very day."

"Opa, I haven't been a little girl for a long time. Tell me. What is it?"

"Your parents. My daughter Irene, and Gray. Their helicopter crash in the Himalayas was no accident."

"No accident!" I backed away from him, my face flushed with anger. "Why? Why haven't you told me they were—what, were they murdered?"

"They were on their way to the NOAA facility when land-cloud seeding was still in use. They were going to figure out if the NOAA simulator could be retrofitted for ocean cloud seeding. By that time, Muller and his crew were involved. There was only one logical conclusion. Purity didn't want more NOAA scientists on site. Purity offered to partner with NOAA and finance an entirely new ocean cloud-seeding simulator. The land-based simulator was mothballed on two of their floors inside the mountain. A new, larger, underground cavity was excavated to house the new science—an ocean-based cloud simulator. They're even calling it the Purity/NOAA facility. It's too dangerous for you to go there."

"That doesn't make sense," I said. "Why would Purity want to help create free rainwater?"

"They don't. They're funding the new simulator so they can be on the inside of every experiment. They will not let it succeed. That's why it's too dangerous. They'll sabotage every experiment. And you with it. Sweetheart, I can't let you go."

I stood my ground and pleaded with him. "Grandfather, even more reason for me to be at that facility. To finish what you—and my parents—spent a lifetime on. Is this Purity thing the reason you've never told me about my mother and father?"

"No, that's not it," he said. "There's nothing anyone of us could've done after it was over. It's so far away over there. I couldn't get a single Senator to investigate, except for Margaret Chambers, and I'm not sure she tried very hard. NOAA tried to help, but nobody's ever been sure of what happened."

"Grandfather, now's the time for us to step up and do what we were meant to do. It's what my parents, it's what your daughter—my mother—would want. Somebody must make another sacrifice, or at least take a stand against these—these—I dropped and shook my head searching for an appropriate name for these despots. That's it. "They're despots, Opa. And you know demand for body water long ago surpassed supply. You predicted it ten years ago! Water companies will not be able to keep up. They're already pushing the production limits of their antiquated systems. It's happening now, Opa! Your own prediction has come true. The CDC is reporting a steady increase in cases of diseases in various areas, mostly third world countries. They claim it's coming from dirty U.S. water."

"What have you learned about dirty water?" he asked.

"Plenty. It's obvious to plenty of us in the science community that enormous bulk shipments coming from water companies who can't get away with selling it here in their own backyard are selling it for cheaper prices to these poorer countries.

"Opa, there are diseases breaking out now. I've read first-hand stories about it. I've seen classified photographs of people with hideous diseases, reportedly caused by nasty water. India for one is buying U.S. water and reporting outbreaks of disease in poor regions. The African government is too afraid to report details for fear they'll be cut off and not receive any water! They claim to have some synthetic iodide that will purify the water. Opa, you have to trust me. I've been on top of this for some time now."

"Grayson, I'm about worn out. These days I just want to sleep. I love my work, but I also love my rest. I used to not care about sleep. I used to have a fire in my belly every morning."

I leaned into him. "And Opa, don't you see? That's me now."

"Yes, I know. I can see it. You've always had your mother's determination." There was a long pause. "What if I say yes? What is your plan?" He wasn't smiling.

"I want you to find that algorithm. And stay in touch. When you figure it out it, you'll need to get it to me. Somehow."

"So, you are bound and determined to go, are you?"

"Yes. I am."

"Okay. When I figure it out, I'll bring it over myself," he said, his voice so much lower now.

"But we need a back-up plan," I said, "in case something goes wrong and you can't get there."

"I'll be there!" It was the first time I'd heard him so gruff.

I twisted my head and gave him my best don't-be-ridiculous smile.

"Very well. Just in case, I'll make a copy and leave it with Albert," he said.

"No, leave it with Jesse, please. Leave it in his brain," I urged him.

There was an awkward hug between us, a quick smile, and I left.

– 7 –

The Way to Gedu

Midnight. The Turnage's small home.

A human figure in the dark appeared on Brooks' porch. The only light was a single moonbeam streaming into the main room. The unrecognizable figure slipped into Brooks' and Anna's bedroom, slowly moving to Brooks' side of the bed.

It reached for Brooks' arm. Brooks' military instincts came alive as he grabbed the arm before it touched him. His eyes opened wide to see Rinku. "Rinku, don't ever sneak up on me like that. I could have—"

"Reverend Turnage," Rinku whispered, "come with me please. Quick."

Brooks looked over at Anna to make sure she was asleep. He eased out of bed to avoid waking her, but she groaned and rolled over to see him sliding out of bed, "What are you doing? It's not even morning. Is that Rinku?"

"I don't know. No. I mean yes, it's Rinku. I don't know what he wants. I'll be back in a minute. Go back to sleep."

Brooks followed Rinku outside.

"Rinku, what is it? What's happened?"

"I know the way to Gedu."

Brooks rubbed sleep from his eyes, "Gedu?"

"The village! The Bhutanese. Coming here," Rinku whispered, his eyes wide enough to let Brooks know that this news was extremely important.

"How do you know the way to their village?"

"I overheard Rashman talking about it at the council meeting just hours ago. I was outside, listening. I know how to find it!"

"Tell me!" said Brooks. "No, write it down. I'll get a pen and—"

"No. I must go. You will never find it alone."

"Rinku. Thank you. But I cannot take you with me. You're too young."

"You, and Anna, and the girls are the only family I have. I must go with you. Together, we will have a better chance to save our village," said a determined eighteen-year-old Rinku.

"Something's not right about this. The girls. Anna. I need for you to stay with them."

"If the Bhutanese come here, they will not care what I have to say. The only way to protect Anna and the girls is for you to stop the Bhutanese from coming here. I don't know how you will do it, but I know you will need me to help with the language!"

Brooks understood the importance of language.

"We need supplies," Brooks said.

"I have them. And hammocks for the trees." Rinku smiled at his own initiative.

Brooks smiled, then turned toward his house. "I have to let Anna know."

"Do you want me to get the supplies and meet you back here?"

"Yes. Give me five minutes."

Adrenaline was running high. Rinku took off as Brooks quietly stepped back inside his bedroom.

"You're going to that other village, aren't you?" asked Anna in the dark.

"Just a few days, baby. You and the girls will be fine. I promise."

"Rinku. Can he stay with us?"

"I'm afraid he has to show me the way."

Brooks leaned down over the bed and kissed Anna on the cheek.

She wrapped her hands around his face. "Promise me you'll be careful."

"I promise," he said with the best smile he could muster, followed by a sincere kiss.

Brooks threw a change of clothes and his Bible in a backpack, then reached for the thermos of water on the nightstand. He quietly swished it around to see how much remained. He determined, *three-fourths full. Good.* He poured half in the canteen, leaving the rest for family. He knew how much sweat he was about to lose, and he would not be any help if he died from dehydration. He turned to see Anna's face perfectly in the moonlight, and he thought, *My God, you are beautiful, Anna.* Then he went to the girls' room. He bent down and kissed each on the cheek and left to wait outside for Rinku. He was in full military mode now. The brief flashback should not have come as a surprise.

> He was hiding behind a wall in Iran, another full moon night, replenishing black paint under his eyes, smearing it in streaks across his cheeks, his weapon strapped over his shoulder. He was about to face an enemy in an unfamiliar city, outnumbered five to one. Fortunately, he was well trained, and thankful for it.

But that was then, and this was May 8, 2036. Rinku interrupted Brooks' out-of-body experience. "I drew this map," he said, pointing. "We go there. Northeast."

"It looks like two days, maybe more," Brooks said.

After two days and a night of foot travel in the rugged low-lying mountains between home and Gedu, Brooks and Rinku were dirty and exhausted. On the second night they again climbed fifteen feet up in dead tree limbs and hung their hammocks side by side.

A fly landed on Rinku's nose.

Slap.

"What was that?" asked Brooks.

"A fly."

"You get him?"

"No, sir. But I will next time."

Brooks stretched and enjoyed a good yawn. "Ahhhh, this won't be so bad for a good night's sleep."

"Maybe for you," Rinku said. "I hate sleeping in these high rises."

"Hi-rise?" Brooks chuckled. "Where did you learn that?"

"The missionary before you. He liked to sleep in trees sometimes. He called it his 'high rise'."

Brooks harrumphed at the thought of a missionary missing big city life.

"He taught me about Jesus," Rinku offered.

"Then he was a good man."

"The best," said Rinku. "He taught me Shakespeare, too."

"Shakespeare?"

"Tis better to stay away from a foolish man, for you will not find knowledge on his lips," Rinku recited.

"Ahhh, good," said Brooks. "But I believe that's from Proverbs."

Brooks turned to look at Rinku. Brooks' eyes turned to mist. Rinku just smiled. They had built an indelible bond.

"I guess now's as good a time as any," Brooks said, as he slowly lowered his head to his chest.

"You going to sleep now?" asked Rinku.

No reaction from Brooks for ten seconds. Rinku thought Brooks had fallen asleep.

Rinku tried to check his map again by the light of the moon. "Yep, looks like we're there, accordion to my map."

Brooks raised his head and laughed. "Accordion? Let me see that map again." Using his small flashlight, he looked closely at it. "I think you're right. Time to get some sleep."

"Accordion," Brooks whispered to himself with a little smile. They studied the stars and wondered what tomorrow might bring. The village wasn't far away. Brooks was more concerned for their safety than Rinku. Brooks said a lengthy prayer. They fell asleep.

– 8 –

Grayson's Application

I WAS A BIT NERVOUS WALKING INTO Purity Worldwide's Headquarters in Arlington, Virginia. I'd already been through the prelims with their HR department, taken the two days of tests, and was told the CEO wanted to meet with me. HR said I was more than qualified, and evidently relayed my test results to Mr. Steve Muller. I wasn't sure if he wanted to meet me because he knew I was Julius Schwarzkopf's granddaughter, or if my test results were so good, he wanted to personally hire me. Turns out it was both, and then some.

I was ushered through a private entrance with two sets of frosted glass double doors, greeted with surrounding blue lights that flashed twice. I knew it was a germ killer. *So, he's also a germaphobe. Most super-rich people are.* I was thankful I didn't have to enter through the employee side with all the stuffed animals I'd heard about.

Once inside his giant office I spotted two sets of sitting areas with sofas and thick leather chairs, a conference room table for nine, and to my right an aqua blue aquarium made into an executive desk. Several computer monitors were neatly arranged atop the aquarium. Blue light columns shot up through the desktop.

The executive chair was turned around facing away from me. A man's legs were propped up on a credenza. The aquarium desk with the beautiful fish was impossible to ignore.

His assistant left me standing there. The man hung up his phone, twirled around in his chair, and bolted up to greet me. He was agile, fit, in his early fifties or late forties. He was wearing a black turtleneck under a cashmere multi-colored pullover and tight jeans. He walked around from behind the huge desk, smiling at me, and when he was close, he casually reached out to shake my hand.

He pointed to one of two chairs facing his desk. "Please, have a seat." He retreated to his executive chair still looking at me. He glanced down at the screen on his desk and punched in a key.

I looked down at all the fish swimming from one end of the desk to the other.

"Grayson Fields," he said. (it was more like an announcement) "Excuse me, *Dr.* Grayson Fields, I didn't know scientists with your brains could be so good-looking. I hope you'll excuse my straightforwardness."

"I'll take it as a compliment."

"Good. We're off to a good start."

I looked back at the fish. "This is unusual." I tried my best not to cringe, knowing that any second some piranha would be let loose and devour the entire school of glorious fish.

"What? The fact that I have an aquarium for a desk?"

"Well, I suppose it does go with the title the press has given you."

"Oh, you mean *Poseidon,* and all that crap. It doesn't bother me. They can call me whatever they want."

I knew that was a lie. His reputation for narcissism preceded him. "Your fish are beautiful. Fresh water, right?"

"Yes, body water." He touched a key on one of the keyboards and bubbles began to grow stronger from the bottom of the aquarium. I twitched and shifted in my chair. I watched as one small fish came up from a tube into the aquarium. I could sense Muller looking at me, not the fish.

"Dr. Fields, I'm introducing you to the most beautiful fish known to man, or at least to me. It's small so you'll have to look closely."

I bent forward and watched as this one-inch-long, pale silver fish with tints of yellow on its body and fins top and bottom almost as large as the fish itself, yellow gills, a large yellow dorsal fin, and bulbous blue eyes passed by me two feet away.

"It's beautiful, but is it dangerous?" I asked.

"It's a Spotted Blue-eye from Australia, or maybe New Guinea, nobody knows for sure," he said. "And no, she's not dangerous. I like her because of her beautiful blue eyes; and she reproduces quite easily."

I felt relieved, thankful he hadn't sprung any flesh-eating display of power on me. "How do you know she's female?"

"Because my aqua people tell me when my girls lay eggs in their new world." He punched another key and said, "Dr.

Fields, we do have an important opening here at Purity for someone with exactly your skills. And education."

"Good. Thank you, I'm excited. Can you tell me more about the position?"

"First, at this level, I always insist on having dinner with every recruit—male or female—before hiring. I'm sure you understand. We just need to make sure you know what that little copper bowl in front your plate is for, and other important graces that can endear our company to our clients and other important people."

Without forethought or common sense, I held up my middle finger, "Oh, you mean the finger bowl."

He laughed immediately. I followed suit. Mine was fake. He looked closely at the little Spotted Blue-eye, and said softly, "She kinda reminds me of . . . oh, well, never mind."

"Of me?" I asked nonchalantly.

He looked up at me. "No, I was going to say my mother." He stood and walked to one end of the credenza behind him. He pointed to a photo of an elegant-looking woman on his credenza. I had to stand and look over his desk to see the photo. That's when I saw my photo on one of his computer screens.

"She was with us here at Purity when I founded the company," he said. "God bless her soul."

"She's beautiful," I said, looking back at her photograph. "But I thought your father started Purity."

"He did. But we never got along. In fact, not at all."

"Please forgive me. I apologize. I had no idea about your relationship with your parents. How long ago?"

"Long ago? For what?"

"Your mother. How long ago did she—did she pass?"

He looked back at her image. "Five years, now. I admit, I still miss her."

So, this must be part of his 'sweet-mother' hustle routine.

"Did you say something about dinner, Mr. Muller?" I said in as sexy a voice as I could conjure up given that my stomach was tied in knots.

"Please, call me Steve. We're all on a first name basis here." Which I doubted seriously but turned on the biggest smile I could muster anyway.

"Then I suppose you should call me *Dr.* Grayson?"

He laughed.

I shook my head a tad and raised my eyebrows to show him a bit of confusion on my part. "So—do you have any technical or other questions you need to ask?"

I thought for sure he'd have the sex question on his mind. *Will he bring it up?*

"Oh, you mean about the position?" he asked. "No. No need for any more questions. I've read all about you. And I like what I've seen so far."

Leaning forward, I asked, "So when will you tell me about the position? What does it entail? If I accept, would I be relocated?"

"At dinner. I'll tell you then. I'll have my assistant contact you once she finds an opening in my calendar."

"And that's all? I'm free to go?" (As soon as I said "free to go" I regretted its sounding too much like he had an upper hand.)

"Grayson, you'll always be free to go."

I received a call the next day from his assistant. "A driver will pick you up tomorrow night at seven sharp. Dinner will be at Le Café Descartes on Reservoir Road. Dress appropriately."

Eighteen hours later I was wearing a stunning new dress with plenty of cleavage.

Here I was, sitting opposite Steve Muller, both of us eating speckled sea trout almandine and drinking a wine I'd never heard of and wouldn't ever be able to afford anyway.

After an hour Mr. Muller was still the perfect host and dinner companion any woman would love to have. He told funny stories, none of them ugly, dirty, or gruesome. And not much bragging. At least not at first.

I was even becoming a decent actress, responding with eye contact to his every word. Still, I wanted to know what this Purity job was about. I had to remind myself that my mission was to push for a chance to work with the simulator at the NOAA facility in India.

After more small talk, the avarice embedded in his soul began to manifest itself. I could sense it like sour wine. He gradually became less sensitive, more focused on money than I cared to hear. He began telling me all about his worldwide operation and successes. He was on his fourth glass of wine.

The waiter took the empty entree plates.

"Mr. Muller, will you and the lady be looking at a dessert menu tonight?"

"No thank you. We don't eat sweets," he said. We smiled at each other—mine was forced. I'm sure his was, too.

"Very good, sir," said the waiter, and walked away.

"Don't ever walk away from me. Until I dismiss you."

The waiter turned and apologized profusely. I don't remember a word he said. I could only hold my head down embarrassed for the man. "Mr. Muller—" I started to say.

"Please, again, call me Steve."

I offered him a quick, tiny smile. "Steve, I don't mean to seem impertinent or presumptive, but could we talk about the position you have in mind for me at Purity?"

He didn't hesitate. "How much do you know about our facility in India, in the Himalayas?"

Bam! He put it right out there. I was taken aback. I wasn't expecting him to bring up the NOAA facility. That was going to be *my* topic. I don't believe he detected any flinch from me.

"I know a little. For instance, NOAA built a land-based cloud-seeding facility over there about ten years ago as I recall. All those clouds, they had plenty to experiment with, right?"

He nodded. "Exactly. Go on."

"And I've heard through Georgetown sources that it's being converted into some sort of giant oceanic cloud-seeding simulator, as if ocean cloud seeding is even possible." I chuckled at the pomposity of such a thing. I wanted him to think I didn't believe in the theory. *Maybe there's a chance he will believe me and try to use me to discredit the experiments.*

"Excellent. You should realize by now why I'm hiring you. I'd like for you to go over and be our number two scientist on the team. You're certainly qualified for it."

"I'm flattered." I gave him a big smile. "Who's number one?" I expected him to say Dr. Julius Schwarzkopf, your grandfather.

"Dr. Bernard Loren, the renowned scientist who knows more about cloud seeding than anyone in the world; maybe with the exception of Dr. Julius Schwarzkopf—(he paused; his eyes twinkled as if he were about to reveal something important)—your grandfather."

He does know who I am!

"So, why wouldn't you want my grandfather, instead of me?" I asked.

"Because I think he's too old and set in his ways. I figure you know how to coax ocean storms back over land as well as he does. After all, that's what he's having trouble with at his own little simulator, isn't he?"

"Yes, he is having trouble; and I believe it's because he's overloading silver iodides in the mixture with the iron oxides. (That wasn't true, but I wanted Muller to think I knew what the problem was.) I continued, "And

Grandfather—whew, what a brilliant, but stubborn man. He won't listen to me." ('Course, that wasn't true either.)

"Grayson, that's why we need you—that's why the world needs you. The simulator and the spray machines and sun lamp are like nothing you could ever imagine. Hell, even I don't understand how it all works." He cast a broad smile.

"Steve, I have just one question. If oceanic cloud seeding is successful, surely, you'd be put out of business. I mean why would the world need your filtration plants if we can create free rainwater?"

He felt compelled to put an arm around my shoulder, "Oh, but you don't really know me, Grayson; I'm an environmentalist at heart. I've made billions of dollars, true; but I've also given away billions to environmental causes. Where do you think your mother got the money to clean up the Chesapeake Bay?"

I was taken aback. "You knew my mother?"

"Of course, everybody knew your mother; and your father. God rest their souls."

I squeezed the stem of my wine glass so hard I had to stop before it broke. I leaned into his face. "I can't believe you're just now telling me this, about my mother and father. Then you know more about their deaths than anyone."

"That's not true. I wasn't over there when the accident happened. I heard it was pilot error."

For some odd reason, I was beginning to wonder if he was just a shrewd businessman who might be telling the truth. I leaned back in my chair.

"Grayson, look, I already have more money, yachts, and more planes than I know what to do with. Hell, many countries don't have the wealth I have. I want to create something sustainable forever. I want something good, something worldwide to be my legacy."

"You mean like really trying to make this experiment work? You really want to see free rainwater?"

"Then it won't matter. Congress will nationalize the water industry. Ignorant Washington bureaucrats will handcuff us at every turn. If that happens, I'll quit."

I had a feeling that Muller would never let enough senators get the bill out of committee.

"Our government has screwed up every private industry they've nationalized," he added.

"I agree with you about the bureaucrats. Nevertheless, the best solution for everyone would be the ocean clouds." A thought popped into my head from the pregnant woman living in the vacant lot on Wisconsin Avenue. *Are you the one to make it rain? Am I?*

The waiter appeared. "Mr. Muller, will you be having another bottle of Hemmings, sir?"

"It was an Opus, 2009," Muller said, obviously proud of his wine knowledge, giving me a flirtatious smile. "No. Just put this on my account."

The waiter bowed and retreated.

Muller turned to me. "Grayson, you know I know a great deal about black water and grey water filtration and all that goes with that, but I know very little about these cloud-seeding simulators, especially for ocean cloud seeding. I've seen the one they're building in India, but I don't know much about it. What's your take on these simulators? Do they actually work?"

He's fishing. "Sure. Science has been using simulators for decades. It's just a way to simulate real-world conditions in a miniature environment to find the right algo."

"Algorithm?" he asked.

Even though he acted like he didn't know what an algo was, I knew he knew. "Yes, it's nothing more than a formula to find the right chemical mixture, the correct spray rate for the chemicals, and the dozens of variables that must all line up in the proper sequence to work. How's that for an answer?"

"I'm impressed," he said. "Do you have your tablet with you? Can you show me the formula?"

I almost had a stroke. I could give him what I knew, or I could give him a variation of what I knew. I decided on the latter. I pulled my tablet from my purse and clicked 'til I found one of the formulas. It wasn't the latest one Opa and I had worked with but it would satisfy my purposes this night. I was guessing he was no brilliant chemist.

I spun it around so he could read it.

After a minute he said, "Well, that makes sense, but if you'll forgive me, I believe there's another formula in there you're not telling me about."

"No, not at all. This last part of the formula, here," I said pointing to the tablet, "this is the hard part. We've got to coax ocean clouds over land and get them to drop fresh water. That's the variable in the algorithm we need to solve."

I was glad to know he didn't know his chemistry. My formula was bunk. "We're about to discover those chemicals and the final algorithm," I said.

"So, you are close to solving it?"

"No, I didn't say that. I said we're hoping to discover the necessary chemicals that can induce clouds to move from sea to land."

He didn't probe any further, but I sensed he knew we were close to solving the science. I needed to change the subject, and quickly.

"You've asked why simulators are so important. Think of the alternative. Sending sea planes out to seed ocean clouds and crossing your fingers hoping the chemicals and all the variables will work to produce the desired result is simply ludicrous, Steve. That method is too time consuming and far too expensive. A simulator can cut the time from months or years to weeks. We just need to find the right algorithm. There are just so many variables to consider; and once it's done," I looked down at the napkin, "that same formula with the coaxing part is repeatable. Forever."

"But didn't you recently tell me it was all impossible, can't be done and all that? Plus, seems to me you and your grandfather have been at it for what, five years now? So, maybe you're right. Maybe it can't be done."

"Steve, I believe we both know Congress wants more information on ocean seeding's viability before they vote on either of the two bills in committee."

"Good for you. You're up to date, Grayson. But which is it? Is it impossible or not?"

I noticed another twinge of avarice, like he was pissed at me for jumping subjects to the two bills that would affect his empire more than anything else. I just knew he was lying about being an environmentalist. So, it was time for me to lay on a bigger surprise.

Intent on avoiding his last question, I said, "And guess what? The committee has asked me to give my opinion on the viability of this ocean cloud-seeding dream. There. Crap. Now you know. Doesn't matter any way. I needed to tell you eventually."

"I see," he said, his eyes, homed-in on me. "You may be surprised to know that I've been aware of your sub-committee chats for some time. But, truthfully, I'd have thought the Senators would need your grandfather's credentials to convince them of cloud seeding's usefulness."

"You don't know my Grandfather. He is too old and frail to make such a long trip and especially be anywhere near Dr. Loren."

"*Not* what I heard from Senator Bennett. He says your grandfather is fit and ornery as an old bull. Speaking of bulls, what's his beef with Dr. Loren? Berney is a renowned scientist. Dr. Julius Schwarzkopf should be thrilled to work with such an esteemed fellow scientist." Muller's crystal blue eyes seemed almost transparent, like I could see through them to his piercing inquisitiveness. His entire persona had gone from charming and witty to dark, on the verge of treacherous.

Damn. Where do I go from here? The wine is affecting my brain. I need to be firm. "Senator Bennett doesn't know my grandfather like I do. And besides, I know as much about oceanic cloud seeding as he does." (Another whopper of a necessary fib.)

Ring. Muller's cell phone rang, or was it his watch? He glanced at his watch and asked to be excused.

"Of course," I said.

As he walked off to talk in private, I let out a long overdue sigh. I wasn't sure how long I could match wits with this man. And never in my wildest dreams would I want Grandfather to be in that facility with Muller and "Berney" Loren, a scientist Grandfather detested. Besides, I knew from the beginning that Muller and Loren were going to sabotage the experiments. If ocean cloud seeding couldn't work, and with Muller's Senator puppets in his pocket, then none of the water companies would be nationalized; and the bill that would give them even more pricing control would likely pass. It's rumored that even POTUS is a close friend of Muller.

When Muller returned, I wasn't sure if he bought my line or not about knowing as much as Grandfather. But he was ready to call it a night. He gracefully wiped his mouth, folded the napkin, and placed it beside his plate. He looked at me with those tender, harmless eyes again. He raised the last bit of his wine and nodded for me to do the same. We touched glasses.

"Grayson, I'm not sure if you've read or heard about this, but I'm receiving *TIME*'s Person of the Year Award in two weeks at the Kennedy Center. I don't have a companion, so, well, I'd like for you to join me."

Bull. I'll bet he has a date. He'll just dump her because he wants me to start trusting him.

"I would be honored." We clinked glasses again and drank the last drop, both of us looking into each other's eyes.

He placed his glass on the table and said, "I apologize, but I must go. I'm meeting with Senator Bennett for drinks. I turned him down for dinner tonight, so, well, you know. I'll have a driver pick you up within two minutes." He even looked apologetic, sheepish eyes and all. He stood to leave. "But look, for the *TIME* ceremony, I'll send a driver to pick you up at 6:30, on the 23rd. Okay?"

I smiled. "Sure, I look forward to it. And Steve—" I cast a broad sexy smile, "thank you for dinner tonight. It was lovely. Oh, and what should I wear for the dinner?"

"Your sexiest black dress. I enjoyed our discussions, too. I look forward to seeing you again."

As he left the restaurant and I waited for my ride, I wondered how much Muller actually knew about my parents' disappearance.

— 9 —

GRAY FIELDS' TESTIMONY

IT WAS APRIL 2024 WHEN MY father testified before the Senate's Worldwide Sustainability of Water Resources Committee. At that time, the Senate preferred to have NOAA scientists testify.

My mother Irene would have testified, too, if it weren't for her fear of public speaking. Two days prior, she came down with a case of "fake" laryngitis. I was disappointed that both of my parents wouldn't be testifying. Still, one of them sitting before a Senate committee in such a historical place was exciting enough for a fifteen-year-old. (Little did I know then that one day I'd be testifying before the subcommittee, only with different Senators grandstanding for the same attention.)

I sat next to Mother, directly behind Father. I was hooked on the proceedings, taking notes and recording the testimony on my iPad—which turned out to be a poor choice for picking up my father's soft voice in such a large space. Recording from behind him didn't help either.

The Senate committee also heard persuasive testimony from the ultra-slick water executives and their guarantees of "safe water for all" diatribes. We attended those as well.

My assessment of those public hearings can be encapsulated in a few sentences: Maybe the common practice of bribes is what persuaded the committee members to act so favorably towards the slick lies of the water barons. No one will know for sure.

The committee made its recommendations to the Senate. "All treatment and distribution rights of body water should be a choice, voted on by local citizens. Municipalities would be required to hold referendums so the public could vote to either keep water supply in the hands of local government or turn it over to private companies for treatment and distribution."

This was the beginning of the "gold rush," or more aptly described as the "water rush" the privates so desperately wanted. To get hold of municipalities' solid waste ponds would mean billions, trillions in revenues for these black water barons.

The water companies began drowning the public with media coverage and advertising messages claiming that municipalities knew little to nothing about turning black water into something pure and pristine. They were probably right about that in most cases. Few cities had the budget to build their own plants and employ the scientists "needed to purify raw sewage."

They campaigned, not only to win over the local municipalities, but to beat each other for the contracts. The privates won nearly all the council votes in the thousands of municipalities across the country. And Purity, behind "Big Tom" Muller's blackmail *squeeze* strategy, won 60 percent of those.

It's how so many city council members were able to sock away tidy sums of money. Taking bribes became as routine as popping open a beer in their Lazy Boys. Meanwhile, the privates banked on a long-term money-making scheme of payoffs worth more than a few beers.

The domestic water rush quickly morphed into a grab for world dominance. Most of the 195 countries put their cities up for play. All the water distributors had to do was secure as many of those cities' rights as possible to sift through their waste ponds and turn their black water sewage into perfectly safe body water. And the most ruthless, the craftiest deal maker of all to come out on top of the mountain was "Big Tom" Muller.

That sinister award still belongs to Purity Worldwide because its founder had planned for the day and knew more about the value of political favors and exorbitant bribes to win his worldwide market share. That process only took four years to complete. After the initial "water rush," the rest of Muller's time was spent staying on top and acquiring favor in Washington, while gradually raising prices of his water.

"After all," 'Big Tom' Muller once told a major news outlet, "when we create something out of nothing it belongs to us to do with as we please."

I was only fifteen when I learned all this during those water baron testimonies. Trust me, I was a nerd, a copious note-taker.

A week later, I received a call from Senator Margaret Chambers asking me to testify before their Water Resources Sustainability Sub-Committee.

"Why not my grandfather? He's certainly more qualified than I am."

"He has testified before. And quite frankly, Dr. Fields, he's not been very helpful. We even have archived testimony from your father, Gray Fields. It was several years ago, but much has surely changed since then. We'd rather have your cooperation voluntarily than subpoena you. Besides, Dr. Fields, this is not like those Senate or House committees you see on TV. We are a small sub-committee, and we meet without any press. We simply want to understand more about cloud seeding and any new developments so we can pass our inquiries on to the full committee. We're meeting with several individuals, including officials from the water companies."

"I assume I'll be able to have my attorney present?"

"Yes, and Pete Gerritsen, head of NOAA, will also appear with you. As you're probably aware, the Senate has some important upcoming votes regarding nationalizing these resources and other private company issues. And there's another issue. If you are indeed going to be at our NOAA facility in the Himalayas using our government's simulator, we have oversight as to who can and who cannot be involved in the project. So, we are also vetting you for this assignment. I'm sure you understand."

"How did you know I was going to the NOAA facility? I only found out myself three days ago."

"Dr. Fields, surely you are aware that our tentacles are far and wide. And, may I remind you we will issue a subpoena unless you volunteer."

I was pissed. But this was no time to argue. "When?"

"July 15, 10 a.m. We will send you further instructions."

I would need to prepare. One night after dinner with Grandfather, I dug out the old recordings from my father's senate hearings. On my screen was a committee of twelve Senators, sitting on their dais.

At the time, I was a teenager. I sat directly behind my father, beside my mother. I turned on my iVid recorder. My father was the sole panelist giving testimony. Unfortunately, some of the recording was sketchy, in the end, inaudible.

Senator Gill asked, "Dr. Fields, you were a presenter and participant in the World Summit on Sustainable Development in August 2002, were you not?"

"Yes, I was."

"And can you tell us who the major players were at the Summit, in your opinion."

"It's not an opinion of mine, but simply the truth: The World Water Council and the World Business Council for Sustainable Development controlled the conference."[1]

"And what was your assessment? In other words, how you would summarize the week?"

"Senator, while the agenda for the Summit was supposed to focus on food security, poverty, and the environment, it turned into a summit on water and sanitation; oh, and it became a profit-making opportunity for the big water companies."

"And how did they manage that?"

"Well, Senator Gill, they basically ran the agenda and the workshops. To be honest, it was a show put on for the water companies."

"It looks to me like they put on a good one, Dr. Fields. I have here before me the total tab spent to wine and dine these influential governmental and municipal bureaucrats in just one of the 5-star hotels used to house the 65,000 delegates. Listen to this: over 80,000 bottles of water, 5,000 oysters, more than 800 pounds of lobster, and I don't need to go on . . . $75 million at one hotel, mind you, was spent to soften up these easy bureaucratic targets. Wow. But it doesn't end there. Listen to this, at the gala opening of the WaterDome, hosted by Nelson Mandela and the Prince of Orange, young people dressed as water drops—or tear drops—flitted from corporate booth to corporate booth to provide the requisite cultural element.[2]

"What do you think the cultural element was, Dr. Fields? (My father didn't wait for a response.) Look, it's obvious to a few of us on the committee that the water companies' motive was to cash in on the lucrative contracts that would open up *if* the summit endorsed private-public partnerships; and they did whatever it took to access this sanction by the United Nations and the 189 governments at the summit. Think about that. One hundred eighty-nine governmental bureaucrats at one summit, being fed caviar and lobster and a mouth full of propaganda. It was the perfect heist opportunity for the water companies."

(I was stunned by her knowledge of what took place at the conference and concerned about my father's answer.)

"Senator, I must say I agree with you. Wholeheartedly. As you know, I've practically given my life to this water dilemma, along with my wife, Irene, here today." Father turned to acknowledge my mother. "Senators, it was evident to us long ago that one drought could propel our world into a downward spiral. I hope I'm not here when that happens."

I fast-forwarded on iVid to Father's closing remarks during his testimony.

"Dr. Fields, do you have any further testimony you would like to share before we adjourn?"

I knew that if the recording carried any clues it was during this part because I had listened to the entire hearings over and over. I studied the video.

My father removed his glasses, placed them on the table and slowly stood up. Placing both palms on the table, he leaned forward, the typical rumpled scientist. Unfortunately, his voice was soft-spoken and harder to hear.

"Honorable Senators," he began, "as you've probably surmised, these are not questions I mull over in my spare time. Matters such as these consume nearly every waking hour for us (he turned to acknowledge my mother again). We have committed three decades of our lives to diligently analyzing our country's, and the world's, long-term water issues." (His voice became sketchy) ". . . countless hours . . . like Sky Seed, American h2O, Spain's . . . can tell you, none of them cares about solving . . . in this for the almighty dollar. American h2O seems especially egregious. And the worst offender" (this part was hardly audible). "There is only one solution." (more inaudible, then) "consequences are irreversible."

Senator Dixon announced himself and spoke, "Dr. Fields, thank you for your time. I believe the committee's opinion is that the solution to our problems is best left to congress."

Father spoke again, his voice so soft I could barely hear what he was saying. I tapped the screen and dragged the video back a few seconds to listen again.

I stopped the recording and retrieved a pair of ear buds from my purse and listened again.

I replayed the last part of the video. With the ear buds I could hear more. *Why hadn't I used these before!*

"They are all in this for the almighty dollar. American h2O is certainly guilty. But the worst offender is Purity Worldwide. They want to make sure oceanic cloud seeding doesn't work."

It was the first time I'd heard my father speak of Purity in that way. I turned the tablet off, tucked it in my bag. I had heard enough. I knew what my father was trying to tell the committee. Purity was definitely the wolf in sheep's clothing.

Now all I had to do was deal with the wolf himself, Steve Muller. But first, I'd need to deal with some stone-faced Senators.

And what I learned from my father's testimony wasn't enough. In the instructions Chair Margaret Chambers sent, I would be given ten minutes for my opening remarks. I had more research to do.

– 10 –

BHUTAN

Sunrise. The second morning of their trek to Gedu.

BROOKS AND RINKU, IN THEIR tree hammocks, were awakened by a noise. "Rinku, shhhh. Hear that?" Brooks whispered.

"Tigers?" Rinku asked.

"No. Vehicles. Coming this way."

Louder. Closer. "Don't move a muscle," Brooks warned.

They watched a jeep, with its machine gun operator stationed in the back, race under their tree, kicking up dust, riding a path over the hill, followed by another jeep, this one hauling a 500-gallon fiberglass tank. Brooks caught a glimpse of an old, faded logo on the tank. *Purity Worldwide.* A third jeep was on the water wagon's tail, also carrying a mounted machine gun manned by an Arab-looking soldier in desert fatigues.

Brooks and Rinku watched the convoy go over the hill and out of sight. They untied their hammocks and dropped down from the tree, into a cocktail of dry air and dust.

"Stay low," cautioned Brooks. "See that bush? Hide behind it."

From their cover on the edge of the hill, they could see the village for the first time and observed the bizarre scene below.

It was barter time. Jewelry, coins, paper money, shawls, handmade baskets—anything the villagers thought had value—were presented to the water lords. The spigot operator took cash first, then pushed others away whose items didn't suit his taste. If he liked what they presented, it would be thrown into a large metal container above the spigot; then he would fill the jugs, pans, gourds, whatever the barterer had, with water from the spigot.

Brooks and Rinku spotted some aggressive villagers pushing their way towards the spigot. Both machine gunners opened fire, killing them instantly. The crowd fell back, screaming in panic. But within seconds, the crowd stepped back over the dead like they never existed, back to the spigot, holding out their items for the water lord to inspect.

It was all over in ten minutes. The gunners fired a few rounds in the air, and the convoy left in a cloud of dust as quickly as it had arrived.

"It's time to negotiate," said Brooks. "May God protect us."

They carefully made their way down to the village. Brooks, in his mind, rehearsed what he would say. *If you won't invade our village, I give you my word I will dig a new well here for you and your village.*

Minutes later, Brooks was sitting on a tree stump, conversing with the village Circle Leader, an elderly man who had more forehead wrinkles than a Shar Pei.

Rinku translated. "Tahir. His name is Tahir."

"Ask him what his name means. 'Tahir'. What's it mean?"

Without even turning to ask Tahir, Rinku gave Brooks the answer, "His name means 'holy one.' He speaks Dzongkha, their native tongue. I am familiar with some of it."

"Ask him if his raiding party has already left for Kupup." Brooks was sweating profusely by now.

Rinku asked, "What do you mean by 'party'?"

"I mean . . . warriors! Ask him when they left for Kupup?" Rinku knew Brooks was nervous, agitated.

In the Bhutanese language, Rinku asked the Circle Leader, "When did your warriors leave for Kupup?"

The holy one looked confused. His deep dark eyes, sunken beneath grey bushy eyebrows and a multi-wrinkled, dark forehead gave him the look of a seer—an oracle. He was the holy one in this village. And Rinku evidently felt some sort of attachment to him.

"Know nothing of what you speak," Tahir said in his native tongue, his piercing eyes focused on Brooks. Brooks began to have a queasy feeling about the village.

Rinku translated what Brooks now knew as obvious: "He knows nothing of what you talk about."

"Ask him if he knows anything about our village, Kupup."

"Holy One, what do you know about the village called Kupup?" Rinku asked Tahir.

"I know nothing of Kupup," he told Rinku in his native tongue.

Rinku said to Brooks, "He doesn't know about Kupup."

Brooks was about to explode. "Doesn't know about Kupup? What? Oh, my God, Rinku. Ask him again. Ask him what his plans are for Kupup."

"Holy One, please tell us what your plans are for Kupup?"

Tahir shook his head.

"Rinku, ask him if this is Gedu?"

Rinku asked.

Tahir said, "This is Chhukha." Then pointed to the south. "Gedu half-day that way."

"Oh, my God, we're in the wrong village!" Brooks jumped up searching for a direction to run. "I've got to get back! Anna. The girls!"

"How can that be?" asked a puzzled Rinku, pulling out his map. "I thought—I thought—"

Rinku's explanation petered out from embarrassment. "It's not your fault, Rinku. But I have to leave. I have to leave *now*."

"No. I can run much faster. You would only slow me down."

"Ohh, my God, please, what have I done?" Brooks shouted, his head back, staring into a cloudless sky.

"I leave now!" said Rinku.

"Wait. Let me think."

Tahir, all the while just watched them, not understanding a word, but able to recognize the panic.

"Reverend Turnage," pleaded Rinku, "I will go in half the time. I will bring them back. I promise."

"You're probably right."

"Let me go," Rinku shouted, jumping up and down.

"But you must just get *them* out. Only them. Wait! I need to write Anna a note."

He changed his mind, "No, just go. Rinku, run hard."

Rinku ran. Tahir's tired, puzzled face showed genuine concern. Brooks paced. Tahir was sympathetic but couldn't communicate. He spoke some in his native tongue, but Brooks couldn't understand a word.

"Is there anyone here who speaks English?" shouted Brooks. "E n g l i s h! Or Hindi! . . . Hindi . . . No Hindi?"

Tahir, for reasons unknown to him, could only stare at Brooks with pity.

Brooks took off in hopes of finding someone who could understand him. On his rounds of the village, he saw several sick people. At the sight of one disfigured man lying outside his hut, Brooks was overcome and turned aside to throw up.

More people were semi-healthy than not, but there were plenty of extremely sick people.

Brooks, shouting now, "Does anyone here speak English? Hindi?" People simply stared at him, puzzled, avoiding his path.

On the trail back to Kupup, Rinku was running like a gazelle, jumping, dodging, skipping through the dead bushes and trees, with no sign of slowing.

Brooks sat beside a tree, his head in his lap, sobbing.

An elderly woman walked up. "You speak English?" she asked.

Brooks slowly unfolded. Their eyes met. Tears in his, cataracts in hers.

"Yes. Please," Brooks said. "You—you speak English?"

"Little."

"Little is better than none," Brooks said.

She turned and walked off. Brooks jumped to his feet. "No, wait. I'm sorry. I must have insult—I must have confused you. You speak English, right?"

"Right on."

Brooks took a moment to appreciate the irony. "Oh, my goodness. Black missionaries?" he muttered under his breath. "What... is... your... name?" he asked.

"Madan. Love, passion."

"It's a beautiful name."

"Brother Green. He give to me."

"Brother Green. Okay. So, there have been missionaries here."

Then, a moment of awkward silence between them.

"Sorry," said Brooks, "my head is sort of spinning right now."

"Right on!" Madan said.

"Look, I don't know where to start. Let me ask this. Is Brother Green here now? Where is Green?"

Madan pointed to a cemetery on the hill, "Green there."

"What happened?"

"Happy-ned?"

"How did he die? How dead?"

"Dis-ease. You see? You like to see?"

She smiled at him. They took off together.

Rinku was running through the hills and ravines of the two-day journey back to Kupup. The hills (locals called them "little mountains") averaged 1,000 feet, the highest was 3,400 feet.

Hours of running and Rinku didn't see the cliff ahead. He tumbled down a 20-foot drop to a dry riverbed. Shaken, but ready to move on, he saw something that sucked more breath from his lungs. Only thirty yards away a half-dozen emaciated tigers were staring directly at him. He froze—just long enough for them to begin stalking him.

He ran down the riverbed ravine and searched frantically for a place to climb out. The tigers were now on a trot to reach him. Rinku was afraid he was done for.

He found some tree limbs hanging down from a steep embankment. He struggled to climb up, grabbing vines, dirt, anything he could get his hands and feet on. The tigers came fast, bunching up below him, snarling at each other for Rinku's dangling legs. Their roars were weaker than normal, but their teeth were close enough for Rinku to feel their hot breath as he kicked at them.

Reaching the top, he collapsed on dry dusty ground, exhausted. He rested; then moved on. Running again. Sweating now more than ever.

Madan and Brooks walked to a make-shift hospital. It was wretched. She took a used facemask from a bucket, gave it to Brooks, and picked one for herself.

"No go near people. Breathe only little," she said.

Brooks thought she might be out of her mind. They walked into what a decimated infirmary ward looked like, only worse. The disfigured faces did not resemble human beings at all. Eyelids had turned inward; faces were covered in boils. Brooks did not speak for at least a minute. He walked through the ward, staring, aghast at the horror, clutching his mask tight to his face.

Brooks ventured a question from under his mask. "Did Green die like this?"

"Yeees," she said.

"Get me out of here."

"You leave now?"

"Yes. Now!"

Outside, Brooks ripped off the mask and puked again. Desperate for water, Brooks said, "Paanee! I need paanee."

"No paanee."

"No paanee?" a desperate Brooks asked.

Madan stepped around the tent and brought back a red, beautiful-looking fruit.

"Phala," she said.

He looked at it for a few seconds, took a bite, and forced a swallow. It was horrible. He spit it out and dry heaved.

– 11 –

My Meeting with Senators

I HAD PREPARED FOR WEEKS, DETERMINED to not let the moment be too big—even though Senator Margaret Chambers said it would be an "informal" meeting with a few Senators, I knew it wasn't going to be easy. I dressed in a new tweed business suit, wore no make-up, and bought the dorkiest glasses I could find. I did not intend for any of the Senators to consider me anything other than a scientist. Pete Gerritsen, head of NOAA approved. "You look like a real science nerd."

"Perfect."

10 a.m., July 15, 2036. To my right was the NOAA attorney, Jed Beasley; Pete Gerritsen sat to my left. Gerritsen knew the meeting location—in the secret Senate basement room. He had been there before.

There were no reporters, no cameras. Chair Chambers welcomed us with polite handshakes and introduced us around the large conference table to Senators Weinhold from New York, Bennett from New Jersey, Salmon from Utah, and Gill from Arizona.

I had met Weinhold and Gill before, only briefly, and I knew of Chair Chambers from Grandfather. He had asked her to investigate the death of my parents, but as he said, "She came back empty-handed, calling it an accident, pilot error."

When we took our seats she said, "Dr. Fields, we are appreciative of your willingness to come today and help us better understand this worldwide drought and what we might do to correct it. Pardon my saying this if you don't mind, but you do look much younger than I imagined. We also welcome your attorney, and of course, Pete Gerritsen from NOAA." She acknowledged and smiled at Pete.

"Pete and I have known each other for a long time and we also value his judgment in this matter. However, we've asked you here today, Dr. Fields, to learn more about a very specific topic of which you are familiar—a topic of great interest to us. We would like to know more about this oceanic cloud seeding and a short explanation of your views on desalination and why it can't solve our problem. I say short explanation, because we've already heard from other scientists on our domestic and international water situation.

"If you'd like to give us a bit more history and background on cloud seeding in general please do so, but as I've asked you on the phone, please keep your opening remarks to ten minutes, after which our Senators may ask you and Mr. Gerritsen a few questions. May we assume you are in agreement with those stipulations?"

She was so cold and unattached I figured she might have been a former spy. A lump formed in my throat. "Yes, ma'am, of course." Pete also agreed.

"Oh, my apologizes, I completely forgot to mention that we represent, minus two Senators who couldn't be here today, the Sub-committee on Water Resource Sustainability. What we learn from our meetings will be vetted and summarized, then passed on to the larger Senate committee. As you've likely read in the news all this material will eventually go before the full Senate for the vote on Senate Bill S5482, the bill regarding the nationalization of water resources and sustainability. You and Mr. Gerritsen may both confer with counsel if you feel the need. This meeting is being recorded only for our benefit. You may begin."

I cleared my throat and smiled. "Thank you, Madam Chair, and all the Senators for having us here today. Our country—no, our world—is at the tipping point for perhaps the most important race for survival in the history of mankind. Forgive me if I sound melodramatic, but I believe we are at the precipice of being left with only a few billion people in less than three years.

"Today, I want to shed as much light on the subject of water sustainability on a global basis and our current condition as you have requested. So first, I ask for your indulgence as I begin with a brief history in terms of water resources, population, and climate, which hopefully will explain how we got to this tipping point. (Thank goodness I had prepared notes for everything I wanted to tell them.) In 1000 A.D., about 300 million people lived on our planet. By the late 18th Century, the Industrial Revolution created a population explosion and one billion tilled the soil, worked in industrial plants, and continued to clear the forests. A mere seventy years later—1920—the population had doubled to two billion; and doubled again to four billion by 1975! We are now at nine billion people. The consensus among social scientists claim that our earth cannot sustain more than ten billion people at most. And that's without any droughts."

I looked up from my notes to see if any of the Senators were listening. They looked like zombies who had heard this story before.

"Senators, our glaciers are vanishing. Our rivers are running dry. Our land is dying. We are on the precipice of distinction. So, what are we doing about it? Let me first take you back to 1949 when America and Israel, followed closely by Africa, discovered how to turn normal clouds into rain clouds. We discovered which chemicals would excite the cloud and give it a power boost so to speak—much like steroids function in our body. Make it stronger, make it produce more crystals to form water molecules. Eventually, the cloud would become saturated and drop its water on land, on crops, on mountains, rivers, everywhere. So that technology has been improved upon around the world. The only problem is that you need clouds over land to turn them in to rain clouds. Although this drought began three years ago, we've been experiencing a gradual decline in surface and subsurface water for many years. And for the last three years we've seen, not just a decline, but by all practical scientific studies, a loss of land clouds. Our lakes, our rivers, even our subsurface water cannot produce enough evaporation to give us the clouds we need to form a decent cloud to even seed."

I stopped to take a sip of water; then continued.

"We are seeing a loss of crops and food source that will only continue. Even as far back as 2000, the earth was failing to feed our six billion inhabitants. Deforestation had created an ephemeral oasis of farmland, followed quickly by a return to desert land. One billion of us lacked safe drinking water, and 2.6 billion lacked basic sanitation. These facts are substantiated.[3] Imagine over 5,000 people in Mumbai India sharing one toilet.[4] It's happening right now.

"By the year 2030, before the drought, half of the world's population faced water shortages. That's when the water wars became serious. At one time our planet included over 300 major rivers crisscrossing frontiers and country borders, which eventually led to disputes, feuds, and wars. These countries that once had life-sustaining rivers running through them, have now lost their lifeline. Aquifers around the globe are already depleted. The worst hit zones around the globe, as you already know, are called 'hot stains.' But the population growth continued to march on, and by 2032, our planet struggled to sustain over *nine* billion people.

"And now, just three years later, we find ourselves in the middle of a multi-year drought." I looked up to emphasize this point.

"Our world population is just beginning to reverse due to starvation at a faster rate than growth has occurred. And, keep in mind, land-locked cities without a port to receive treated water are in the worst condition; they also have no way to desalinate saltwater. So, here we are."

I stopped to glance over at Gerritsen and Beasley. Pete showed a hint of a grin which signified I was on a roll. I looked around the table. But what I saw was the underside of a bunch of noses, utter arrogance, and bored faces.

"Quite frankly, Senators, our world has run out of available clean water. And it is your fault." That got their attention.

"Senators, this is the virtual comet we have so feared, except it's not a rock or a boulder of ice coming at us from outer space, it's our lack of the life-sustaining gift of water. The villages that have water wells are destined to continually protect the little they have. Villages are being overrun by water war lords. It won't be long before it happens here. In fact, it's already happening." I took a breath and glanced down at my notepad.

"Dr. Fields, we hear you. Now tell us where you think we stand regarding a potential solution," said Margaret Chambers.

"Okay, let me start with desalinization. Simply put, and as you have no doubt heard or read many times, it is *not* the solution. For every single liter of clean water produced, desal leaves us with a liter of poisonous by-product pumped right back into the ocean. Yes, a city here or there on the coastline of all countries can use desal as a secondary source but it is not enough for them to survive and desal is certainly not available to land-locked cities.

"This brings me to black water plants. Black water plants have been in existence for decades. I want you all to know that conversion of black water into safe body water has been accepted as proven science since the 1970s. It was NEWater® in Singapore who first developed the technology for converting raw sewage solids into potable, drinking water. I suspect most, if not all of you, are aware that the water you have in front of you was once sewage. Don't worry, I'm sure it's perfectly fine. But the problem with black water is imminent. I'm not sure how much you Senators know about the process, so I can skip this part if you'd like. It's fairly graphic."

"No, by all means, proceed. We've heard other versions, but we'd like to hear yours—for comparison purposes," Chair Chambers added, without blinking.

"In these black water sludge plants, raw sewage flows through large sedimentation tanks, where the solids settle to the bottom, and mechanical skimmers rake off the oil and grease from the top. The sludge left in the tanks is pushed through pipes at the bottom and into secondary treatment tanks, where aerobic biological processes are used to substantially degrade the sewage. The sewage is then treated with chlorine and other purifiers, and under high pressure, forced through a bank of large twelve-meter diameter membranes for more cleaning.

"By 2034, these membranes had become the bane of every water company's existence. If just one of these membranes deteriorates, more sewage solids than allowed by the FDA and the Water Resource Sustainability Cabinet, would enter the various chemical processes and cause problems at the end. A tiny hole can mean deterioration. That liquid from the tiny hole will find its way into 10,000-gallon tanker trucks, labeled as 'PURE', all the way down to six-ounce water bottles, just like the ones you have in front of you."

I took the time to relish this last comment by looking up from my notes to see if any of them were gazing at that bottle in front of them. Gill, Bennett, and Salmon were. Chair Chambers never took her steely eyes off me.

"Senators, this is not new information. One report estimated that in developing countries, water demand will exceed supply by 50 percent in 2035. And so, that prophecy has come true. Except now it has been exasperated by this drought. It's where we are today and I, for one, am convinced that oceanic cloud seeding may be our last and only solution. I'm not saying all black water filtration is tainted. In fact, only a small percen—"

Senator Gill interrupted with a hasty question: "Tell us what the percentage is and who the culprits are that allowed this to happen," his face red with anger. At the time, I didn't know Gill's Arizona was experiencing problems with tainted water.

"Sir, I've read from science journals that the average of all studies done is approximately 4 to 5 percent of these plants are experiencing filtration problems; however, we can't be certain of the number. The smaller operators simply don't report their amount of leakage. But Senator, the bigger problem lies in the near future. The rate of growth for tainted water is growing at 15 percent per year. These numbers have been tracked for three years, so it's sketchy to extrapolate too much."

I placed both hands on the table to emphasize this next point. "But, trust me, by 2040 it's possible that we will see a lot more tainted water coming from plants all over the world. Of course, we'll all do our utmost to make sure that doesn't happen." I cast a broad smile at them all. "I don't know about you, but I'd like to be here for a while."

They all laughed. I'm sure they all wished for the same.

I continued. "As our world's population has grown, so has the demand for body water, as well as AGwater, which we need for food. Agriculture uses 70 percent of all fresh water. The demand just continues to grow. Did you know that it takes 2,900 gallons of body water to produce one pair of blue jeans? How about 53 gallons for one glass of milk? Or 379 for one pound of figs.[5] Senators, I apologize if I'm giving you more details than you want. But it is important to look at facts."

"Tell us what the water companies are doing about the tainted water problem," Gill continued.

Madam Chair quickly looked across the table at Gill and said, "Ray let's allow Dr. Fields to complete her opening remarks if you don't mind and then ask questions. She's doing fine so far."

He nodded. "I apologize to the Chair."

Chambers looked at me and nodded, giving me the go ahead to continue.

I turned from Gill to Chambers, "Senator Gill's question is a nice segue for my next point. As this demand for body water continues to rise to unprecedented levels, most of the large water companies are doing two things. They are investing in new membrane technology and they are continuing to build new plants in more cities. Unfortunately, there will always be the renegade small and medium size operations who will fill tainted water containers headed to poorer countries for a cheaper price."

(Of course, the larger companies were just as guilty. But I knew my last comment would get back to Muller and help build his confidence in me.)

I continued. "There's something else about black water plants. If one of them experiences a problem, say, a gasket in the piping deteriorates and leaks, the plant cannot shut down. The problem must be fixed, under production, using a by-pass pipe, similar to open heart by-pass surgery. If the plant were to shut down, the sludge in the entire system would coagulate and form a thicker solid and lock up the whole operation, bringing it down for months, perhaps forever."

I reached for the water bottle, and this time, I slowly poured some into a clear plastic cup, all the while studying the water as its silver reflections flowed into the cup. I could imagine each of them watching, wondering what might come next.

"As you know (I held up my cup) … this was once sewage. Raw sewage." I paused to let that sink in as I swallowed. "And right now, I'm hoping there's no proverbial fly in this ointment. But what if there is?

"Senators, imagine, seventy or more small to medium size water filtration companies around the globe. Most of these are small with one-city or two-city contracts. They're not nearly as sophisticated as the larger operators. These small operators simply don't have the wherewithal to maintain proper quality control. Even here in the U.S. we are hearing spotty reports of tainted water popping up in various cities. Evidently Senator Gill is experiencing some of these problems in his state." I looked at him. He nodded his approval.

"I sure as hell am," he said.

"Senators, it's happening all over the globe. A group known as *Citizens Against Drinking Sewage* in Toowoomba, Queensland, Australia, has held up plans to recycle sewage water for years. They've been circulating propaganda campaigns against the very thing you are drinking now."[6]

I looked back at the Chair. "Most of the problems seem to be coming from membrane failure. The reputable water distributors claim to have new membrane technology that eliminates any and all bacteria. Frankly, I believe them. But, as someone once told me, 'Even if you put a single small drop of sewage water into a bottle of Dom Perignon, you would've turned a fine champagne into a bottle of sewage.' How much more dangerous if it's more than a drop?"

The Senators looked too complacent. I needed to shake them up; I needed to make them lean forward and listen with rapt attention. That's why I brought the slide show. *If this doesn't put them on high alert for what's about to hit our country and spread like wildfire to the four corners of the globe, nothing will.*

I placed both palms on the table. "I want to take five minutes to show you some images taken at various hot stains around the globe. These images are shocking and may turn your stomach, but I must assume, in your position as Senators you want to know as much as you can about what's going to hit our shores sooner than we can imagine. You'll see sick and dying villagers with boils and what looks like hives covering their bodies. There are other images of people suffering from water-borne diseases. Yes, there's not just one, but many variations of the disease. Look closely at some of the photos and in the background, you'll see water trucks with company logos printed directly on the tanks."

I looked around the room at each Senator but remained on Bennett a tad longer than the others. His already little beady eyes became thin slits. I knew he wouldn't want Purity's name to show up on one of those slides.

I took a moment to study my notes, then addressed the room.

"These diseases—" I stopped and nodded towards my computer handler, "If you would please bring up the photographs."

The screen remained blank. Nothing. Nada. I raised my eyebrows at the computer operator. He turned his palms up and shook his head. I was astonished. *Crap. Someone must've hijacked our computer.*

Senator Weinhold spoke: "Perhaps, you could produce that evidence for us at a later date?"

"Of course," I offered. I didn't know what to say. I was embarrassed. I'd lost my momentum. I stumbled through this next part. "Senators, I hope you realize what's happening. This was working early this morning."

"It's okay, Dr. Fields, these things happen," said Chambers, as she glanced at Senators Weinhold and Bennett. She then told me to proceed.

I was too pissed to look at any notes. I took off on an angry rampage. "I was going to show you photographs of what the water disease is doing to tens of thousands of people in poorer countries. Let me just say this. The WHO reports that contaminated water is implicated in 80 percent of all sickness and disease worldwide. As far back as the early 2000s the number of children dying from diarrhea exceeded the number of people killed in all armed conflicts since the second world war. Every eight seconds, a child dies from drinking dirty water.[7]

"And now the problem has worsened. When the municipalities of *all* nations put their waste ponds up for bid and let the water filtration companies contract with them to turn poop into gold! That's when it became worse.

"The companies began to expand; they became too stretched, and too focused on profits to remedy the enormous demand for body water. Most of them had plant problems. What did they do? They allowed filthy water to be sold at cut-rate prices to poorer countries." I breathed and gained control of my anger. "Still, I believe most of the inferior water is coming from the small operators in foreign countries."

My heart was still racing. "Senators look at that bottle in front of you. Before long, you won't be able to trust what's in it. Is it a Dom Perignon of body water, or is it sewage? Thank you for your time today. I look forward to answering your questions."

My face was now flushed. I was beyond upset that the images of those poor diseased people weren't shown to the Senators and more importantly, to the public.

The chair thanked me and called for a short recess. When the Senators returned, Pete and I answered questions for about an hour. Since none of them were scientists, their questions were softballs for us. "What exactly is an algorithm? How many planes will it take to seed an ocean cloud? What prevents the storm from becoming a hurricane?" I enjoyed answering that one.

"It all depends on the volume of chemicals—and we use several—that are typically dropped into an already formed cumulus or stratiform cloud. It's kinda like a controlled explosion. Within 30 minutes of seeding a cloud, rain should fall. I won't bore you with how the cloud is super-heated, but eventually ice crystals are formed that pass through the heated layers and become freshwater drops. I'll include the science of that in the leave-behind material."

Chair Chambers seemed to be ready to close the meeting down, but I soon found out she had something else in mind.

"Dr. Fields, and Mr. Gerritsen, I personally want to thank you both for being here today and helping us understand the dilemma that faces us, both here and abroad. Now, you mentioned that our best chance of turning this around has to do with coaxing these ocean clouds over land. That is the mission, is that right?"

Pete Gerritsen answered. "Madam Chair that is exactly what we must do in the Himalayas. We must find the proper chemicals, in the right proportions. It will require some brilliance on some people's part. I believe Dr. Fields here is our most qualified scientist to represent NOAA. We have a dozen other scientists on the team, but I've put her in charge from our side. I'm confident you can see from her performance today that she is more than qualified to represent us in what will surely become our country's and the world's most important mission."

"That's not the full extent of my question. I also asked how this coaxing of an ocean cloud is believed to work. I'd like to know that, and I'd like to know what you believe is its probability of success." The whole time she focused in on both at me and Pete Gerritsen.

"If I might add to Mr. Gerritsen's answer," I started. "There are many other variables that must be considered other than just coaxing an ocean cloud over land. Seeding an ocean cloud is much different from seeding a land cloud. But I believe we have solved that issue."

She turned to Gerritsen and said, "Mr. Gerritsen, we assume you'll also be at the facility while these experiments are underway, right?"

"Yes, Senator."

"Then will one of you tell me how it's done and the probability of success?"

I knew she was phishing for my grandfather's theory. She wanted to know what chemical he was using. I turned to Beasley and asked a question. Once he answered me, I was back onto Senator Chambers.

"Well, madam Chair, I'd like to address that. While we appreciate your desire to know how we intend to do it, we hope you respect that the method and chemicals we're using is actually proprietary. If I divulged it to the public, it would no longer be proprietary."

Her eyes widened. For the first time I saw more brilliant blue than I had for the past hour. She leaned forward on the table. "That's not how our committee on sustainability works, Dr. Fields. You are here to answer our questions, no matter how inquisitive or proprietary they might be. And remember, your answers are not for public knowledge." Her eyes were penetrating and menacing.

I knew her "not for public knowledge" was pure BS. Surely, she'd lived in Washington long enough to know nothing is secret, especially fake gossip. I leaned over to Beasley, seeking an answer to Chamber's question.

He gave it to me.

"I'm sorry, Senator Chambers, but that's *not* exactly how it works. As a private company, we are not required to divulge our potentially patentable designs or ideas. Besides, I need to add that none of this has been proven or successfully tried."

"But you are working for a private company, are you not, and you are about to work with our nation's National Oceanic and Atmospheric Administration in a secret facility? Dr. Fields, may I remind you we are the oversight committee for NOAA and our Water Sustainability new branch of government."

"Senator Chambers, I represent a private company as a consultant to the NOAA facility in the Himalayas. That's all."

Senator Bennett broke in, "I was under the impression you were working for Purity. Is that not correct?"

"I'm merely a consultant hired for a short-term project. I work for an LLC named Clouds Above."

('Course I'd never been paid; all I got was room and board at Grandfather's crumbling castle.)

"Senator, if our theory works, fine, if not, there'll have to be another solution to the problem. NOAA believes ours will work; and if successful, believe me, every country in the world will be flooding the skies with our chemical formula."

"Who owns Clouds Above?" asked Salmon from Utah. A question I hadn't expected.

I asked Beasley how I should answer the question. What he said didn't surprise me.

"Senator, it's a limited liability corporation with two managing partners, Dr. Julius Schwarzkopf and Georgetown University. If the formula works, the licensing rights will belong to the managing partners."

"You mentioned formulas. Is there more than one?" Salmon asked.

"Yes, there's a combination of formulas—three are needed for coaxing the cloud over land, and another for increasing a cloud's vapor volume. All of these must work in perfect synchronization. Again, it's why we experiment in a laboratory with an actual simulator. It would simply take too long to find the right algorithm for all these variables. We'd have to send planes into clouds hundreds or thousands of times out over the ocean just testing the various combinations of chemicals."

Chambers asked, "As I've understood it, companies like h2O, Avanti, Purity, and others have shifted their filtration plants from grey water to black water. Can you explain, Dr. Fields, why they would do that?"

"They haven't shifted anything. Senator, they've been into black water filtration for decades, as I'm sure you are aware. The grey water recycling systems now inside most all residences and businesses plateaued three years ago. The new emphasis has been on black water conversion. It's all about the money.

"Our world is now faced with having only a finite amount of fresh water. There has been no rainwater or snow melt to speak of for four years. We are forced to recycle what we have.

"We all have a grey water filtration in our homes. That's all fine for bathing, and some to replenish your toilet, but in the end, it's a finite source that must be recycled. It cannot be turned into clean body water, or else we'd be without any water of any kind. Our clean, body water must come from another source. Black water. It's why our Water Resource Sustainability cabinet and Congress mandated that we all use black water filtering for body water. In essence, we are recycling all the available water on earth for all our needs. Water is being recycled and recycled again. As I've said, we simply don't have enough new water from land clouds or snow or glaciers to refill our lakes, our aquifers, or meet our agricultural needs."

Senator Weinhold dropped his clasped hands on the table, obviously frustrated.

"Dr. Fields, why didn't scientists see this coming earlier?"

I wanted to slap him, wake him up to reality. With as much disdain as I could muster, "Sir, our earth seems to have given up producing fresh water. It has been evident for some time. It's Congress that didn't act earlier." I didn't care what he thought. "And I firmly believe oceanic cloud seeding is our only answer. Again, remember, several studies show that 90 percent of ocean clouds dump their clean water right back into saltwater. Only 10 percent reaches land.

"Water has no longer simply become a limited resource, Senators, it has become a recycled object more valuable than plutonium." The arrogance on their faces had now been replaced with concerned eyebrows. Nobody moved for five seconds, and that's an eternity in a moment like this.

Chambers asked, "Dr. Fields, let me ask you—are you in favor of private industry being in control of our water filtration and distribution, or is it your opinion that the industry should be nationalized?"

Finally, here's what the meeting is all about. They want to know whether they should vote for or against nationalization. *It's all politics.*

"Senator, I believe we are always better served by reputable and trustworthy private enterprise handling complicated matters such as water sustainability. I have little faith in our government's ability to handle it correctly. Bureaucrats have no incentive to make things work better. They have an excellent track record of mucking things up. However, the operative phrase in all of this is 'reputable and trustworthy private enterprise.' We will need extensive oversight of the quality control of every plant around the globe. That's where Mr. Gerritsen, his team, and the newly-revamped World Health Organization come into play."

"Mr. Gerritsen, do you agree with Dr. Fields?" asked Chambers.

"I do. Right now, private companies treating black water know more about how to do it than any government. Is there bribery in the industry? Yes. Are their bad actors in the industry? Yes. But rather than increase regulations, why don't we first find a way to bring rainwater back to our planet then let's oversee its success with regulations that make sense."

"And your idea of how to do that?" Weinhold asked.

Pete didn't hesitate. "It starts with our ability to successfully create and replicate oceanic cloud seeding. It's our best chance. If we can do that, I have no problems with our government using the Air Wing of NOAA to oversee the entire seeding operation. QC scientists will need to be stationed at every plant."

Although Muller wouldn't be happy hearing this from Gerritsen, it's probably what he would expect Gerritsen to say. *Maybe Gerritsen should think about bringing a bodyguard to the facility. He may need one.*

Chambers muttered, "Humph." She looked around the table at her colleagues, "Do you have any more questions for Dr. Fields or Mr. Gerritsen?"

Head shakes all around, until her inquiry reached Senator Gill. "I think Dr. Fields and Mr. Gerritsen have given us plenty of information and are to be congratulated for their input today." Heads nodded. "I have but one more question," he added. "When will we know if oceanic cloud seeding is successful or not?"

This was best handled by Pete Gerritsen. "If all goes as planned, I believe you can expect to receive a report within a month, maybe two." I figuratively crossed my fingers behind my back and thought, *all they have to do is follow the money.*

The meeting ended on a swift thank you from Chair Chambers and we dispersed. Gerritsen, Beasley, and I headed for the nearest bar. We were exhausted. After a few French martinis I decided to drop by my favorite neighborhood sushi bar and have something to eat. I never figured on it being the scariest night of my life.

– 12 –

Follow the Money

The New York Commodity Exchange.

IT'S AN EERIE BUILDING NOW. Devoid of people, except for a few lab-coated engineers who monitor the mainframes for security and constant back-up. The old days of pit traders elbowing each other for position and yelling out their buy/sell orders were long gone. For years the trading has been done electronically in a secret, underground bomb-proof New Jersey location. All trades come straight from the seller's computer to the Jersey mainframe, where the sale offer is instantly posted to all licensed traders. Buyers enter their IDs, the buy order, and press send. The deal is consummated.

One Index that's been on the "big board" is the ISE Water Index. Operating since 2006, the International Securities Exchange Water Index represents companies specializing in water distribution, water filtration, flow technology, and water-related solutions. Having over 120 stocks, the exchange offered an indexed price of the combined companies, but by 2014, individual stocks were the new equivalent of the tech boom decades earlier. The

Top 5 Watch List of publicly traded "darlings" were Energy Recovery, American h2O, Flow International, Tetra Tech, and Purity Worldwide. By 2020, water futures were where the gamblers hung out. It's where the real money was made. And lost.

While Purity was technically publicly traded, its founder, Steve Muller, had managed to retain enormous holdings in the company—24.7 percent of a three-trillion-dollar market cap company, making Muller the wealthiest human in our known galaxy. Personal net worth? At least one-and-a-half trillion. It looks like this: $1,500,000,000,000. He could buy countries. The stock (PURW) traded for 11,000 per single share. Muller was the darling of the darlings. Or, as the Wall Street Net Journal reporter dubbed him, "Poseidon, *the water god.*"

Steve Muller was destined to carry this mantel, having been born into the home of Purity Worldwide's founder, "Big Tom" Muller, who was so hated and loathed by nearly every person who knew him for his unsavory business tactics that his funeral was reserved for family only. And not many of them bothered.

Following young Steve's Harvard education, he immediately entered the most exciting world he had ever known and studied in school—the fast-paced, cash-flowing, money-making world of water. It was the only industry that had replaced the economic boon of any era in history.

And most of it took place on the trading floor of the World Commodity Exchange.

At the top of that heap was Seinfeld & White, a highly respected brokerage house, with 5,000 brokers in Orlando on the Disney property. Aggressive sales mavens, money makers. Men, women; each ate what they killed. Their kids attended K-12 in the Magic Kingdom. Everyone ate high on the hog—better than 99.9 percent of the world. And why? Because Florida still had no income tax.

For the broker, the client list is and always has been the Holy Grail. Phillip Percy, Grayson's former husband, spent six years developing his blue-chip list of wealthy clients. He recently spent an hour convincing a new prospect (the Investment Committee Chairman of Bio Xquities) that they should load up on Purity and flip it for a quick profit. Bio Xquities had a reputation as risk-takers. Percy practically guaranteed a 12 percent return. He had been studying and following Purity stock ever since Steve Muller's presentation at the International Water Symposium where the silky-smooth Muller impressed everyone with his assessment of the world's water situation and Purity's strategy to capture even more market share. It was at this road show presentation that Muller unveiled Purity's new membrane technology for rapid wastewater filtration.

But it was Muller's hint that congress was going to pass bill number S5482—the bill which would give water companies more latitude on pricing—that sent money managers and brokers like Percy into hyper-palpitations.

Early Monday morning Percy was working next to his competitor-broker-friend, Jimmy Dawkins.

"Hey dumb ass," Percy said to Dawkins, "guess what? I just bought 21,000 shares of Purity. Yes! Yes! My commission will be over two million." Phillip Percy threw his head back and laughed so loud half the floor turned to look.

"Phillip," Dawkins said, "didn't you see this morning's report on Purity's treatment facility? Here, look." Dawkins brought it up on his screen.

Percy craned over to look at Jimmy's terminal. An article about some problem with Purity's treatment plants appeared on the screen.

"Oh, no," Percy murmured. "What's this supposed to mean? It says their plant in Philadelphia is having problems. And there're some rumors about their Sacramento plant."

"It means your client is going to be super pissed off. Just keep looking at the big board. The stock's going to start dropping unless the company can put some freakin' strong perfume on this."

Percy dropped his head onto his keyboard and slammed his fist into the desk. "My timing. My timing sucked. I should have checked to see if there was any breaking news."

This was the financial center of the world in 2036.

– 13 –

Troubles Everywhere

DR. BERNARD LOREN, MULLER'S prized scientist, is on board Purity's helicopter, with his assistant, leaving the D.C. skyline behind, headed west to Philadelphia. Loren is forty-one, polished, balding with round specs—an "in charge" kind of guy. Always on his cell. Always being a bully.

"Simpson, this is Dr. Loren. I'm on my way there right now. How bad is it?"

"Sir, I'm not going to BS you. The suspended growth systems in the last six shipments were not floced correctly. It's a—"

"Dammit, Simpson!" Loren interrupted. "Why wasn't maintenance done on schedule?"

"Sir, we did the maintenance. That's when we noticed the foul odor coming from the wash water and the water spots and residue on the pumps, even after we had cleaned them with robotics," Simpson quickly said.

"Well, it better be fixed asap."

"Four days, tops."

"Not acceptable. Word is already on the street that we're having problems. Philadelphia will be a PR disaster with congress and the market. I'll be there in an hour. I want it fixed by end of day tomorrow."

"Sir, we have to run a by-pass around the joint where the biggest problem is happening. And sir, what should we do about the six shipments?"

Loren, calmer now, "What shipments?"

"The ones on our trucks headed for the port."

"You mean they came out of the batch that's screwed up?"

"I'm afraid so, sir." Should we dispose in the pond?"

"Who's the buyer?"

"I'll look it up; one second."

Simpson came back online, "Sir, it's headed for Senegal. Twenty million in gold bullion."

Loren paused, lowered his head to think. Rubbed his eyebrows, contemplating, and said, "Let it go."

At the same time, Muller was calling Loren. "Bernard, what happened? Heard you're headed to Philly."

"We have a little problem within the secondary treatment process. I'll get it fixed."

"Good. Guess what? Bio Xquities just bought 430 million dollars of our stock!"

"They may live to regret that decision if we can't get this fixed. Steve, we also have a problem in Sacramento. I think it'll be the same as here. We're working to get it fixed. But, Steve, we're just pushing too much black water through the membranes."

"I know you can work it out, Berney. Look, I was hoping you could come down to D.C. for the *TIME* dinner. I have a surprise for you."

"I'll do my best, Steve. First, I have to fix this problem in Philly."

They hung up.

Loren turned to his assistant. "He sure sounds giddy, even unconcerned about the plant problems. I guess he knows I'll fix it. I always do. Get me the firm in Philly. The PR firm. Get their CEO on the phone. What's his name?"

"It's Damon, sir. Matthew Damon."

"Get him on the phone."

A minute later, Loren's assistant says, "I have the PR firm on the phone, sir."

Loren took his assistant's phone. "This is Dr. Loren. Is this Damon?"

"No, sir. I'm Mr. Damon's assistant. How may I help you?"

Loren cupped the phone with his palm and turned to his assistant then gritted his teeth and unleashed this venomous chastisement: "Why the crap did you bother me unless the CEO was on the line?"

"I'm so sorry, sir." His assistant began shaking.

"And to think, I pay you $200,000. At least you know what to do at night."

He turned his attention back to the phone, "Would you kindly get Mr. Damon on the phone for me?"

"I'm sorry sir, but he's not in the office at this time." Followed by a long pause.

"Okay, then, tell you what. Have him call me within an hour if he wants to keep our business. It's only worth, let me see, (his assistant holds up five fingers) yes, five-million dollars."

"Oh, heavens, I'm—"

Loren presses "End Call" then uses voice command, "Call Sacramento plant."

Three rings.

"Good morning, Purity Sacramento. How may I help you?"

"Rob Schmaltz, please. This is Dr. Loren."

"Mr. Loren, is he expecting your call?"

Loren pauses. "I doubt it. Why don't you tell him that Dr. Bernard Loren, President of Purity Worldwide, is on the phone, and if he—and you—would like to keep your cushy jobs he better get his butt on this call in thirty seconds. Understood?"

"Mr. Loren! I mean Dr. Loren, I am so sorry, I didn't—we never—"

"Miss—?"

"Jennings. Jane Jennings. Please don't hold this against me."

"Look, Jane. You're probably new, but right now you're wasting my time. If you'd just get Mr. Schmaltz on the line, I'd appreciate it."

"Yes, sir. Right away."

(Pause.)

Loren mumbles, "Schmaltz is too thick-headed to realize what's going on."

"Dr. Loren! What a pleasant surprise. It's been a year since we met at the annual—"

"Schmaltz! Shut up and listen."

Schmaltz' trembling voice now became obvious. "Yes, sir, I'm sure you just want me to fill you in, right?"

"That's right, what the crap is going on out there? Don't tell me we're having a secondary filtration problem!"

"Sir, I believe we are. It started last night. Frankly, sir, our production schedules are a bit above our capacity to—"

"To what, produce what we need? I'm well aware of what goes on without maintenance. Your job was to inform us. If you had taken the time to look at our Purity engineering posts yesterday, you would've seen that we're having the same freakin' problem in Philadelphia. Not only that, you could have averted this problem by changing out the gasket. Now I imagine the membrane has deteriorated, too."

"My bad, sir. I guess I just didn't—"

"Think," Loren said. "You didn't think; and you're fired. Get your number two on the phone."

"Sir, he's off today," came a deflated voice.

"Tell you what. I'll just be in touch with him later." Loren hung up, turned to his assistant, and said, "The number two is a new guy. I can't remember his name."

"It's Hobbs. Henry Hobbs."

"All I know about him is he likes to play golf. Get his cell number. His butt is probably on some golf course."

Loren had been with Muller at Purity from the beginning. He was Purity's first scientist and was a celebrity in his own right. He knew the operations side of the business backwards and forwards—the collection of grey water and black water, the treatment/processing side of the business, bottling, and shipping.

His biggest headache from the beginning was dealing with municipal government safety regulations. The bureaucrats were his nemesis. But his weapon was the stockpile of Purity money.

The helo pilot's voice came over the intercom: "Sir, we're twenty minutes out from the Philly plant."

"Give me this week's hot stains report," Loren asked his assistant.

She pulled out her tablet, touched one key and handed it to him.

The first report was from Buenos Aires.

Buenos Aires, Argentina
Summary

Water trucks have been hi-jacked by looters, with no compunction about shooting our drivers and guards. Government buildings have been taken over by the revolt. Man or woman, if they look wealthy, they are targets.

Loren stopped reading about Buenos Aires and switched his attention to South Africa.

Cape Town, South Africa
Summary

Population now twenty-five million. Purity market share is growing due to SA-Filtration system breakdowns.

Loren quickly scanned that report, and clicked through several more, looking for something.

"Tell me what you're searching for and perhaps I can help," his assistant said.

"I'm trying to find out if we're having problems in other plants."

"I can help," she said, wanting to please. "I believe I read where there's something similar going on in Scottsdale."

"Scottsdale?" said a surprised Loren. "Let me see the report."

His assistant took the tablet and found Scottsdale.

Scottsdale, Arizona

Loren quickly scanned the brief report. Scottsdale was beginning to have similar overload problems as Philadelphia and Sacramento. Without speaking to anyone in particular, Loren muttered, "Why didn't Steve warn me about this? Ever since he met that woman, he's been in La-La Land. Get me Scottsdale on the line."

"I'm searching for his number right now. And, Dr. Loren," she said, handing him another story from the tablet, "have you seen this latest memo from Senator Bennett on the Sustainability Committee?"

Loren snatched it from her hand and began reading.

"Yep, I'm right," Loren muttered to himself. "Dr. Fields has to go. She told the sub-committee far too much information; here, look at this part, she even told them to be wary of us, Purity! I warned Steve about her, dammit."

His assistant took a call from the number two man at Sacramento, and sure enough, the background noise sounded like he was outside, not in an office. "Yes, this is Henry Hobbs. I'm trying to reach Dr, Loren. Is this his number?"

"Yes, indeed, it is, Mr. Hobbs. Hold on for Dr. Loren."

"Loren here."

"Dr. Loren, Henry Hobbs. I understand you were trying to reach me. I apologize for—"

After hearing birds chirping in the background, Loren interrupted, "No need to apologize Henry. I hope you're beating the crap out of those other schmucks on the golf course."

"Sir, I'm not quite sure I'm following you."

"Are you not on the golf course?"

Loren heard Hobbs cover his phone, but Hobbs' muffled voice came through ever so slightly, "Guys, hold it down, my boss in on the phone."

Louder now, "Dr. Loren, no sir, I'm just in my car on the way back to the office. Would you rather we talk then?"

"No not really. Now is just fine."

"Do you have some news or a question for me?" Hobbs asked. Loren could smell a nervous voice like a snake in the dark senses its prey.

"Not really, I know everything I need to know. I was just checking on your character, and I got my answer. Have a good rest of your day on the golf course, Mr. Hobbs."

Loren held the phone out to his assistant, with these words, "Tell HR to start searching for a new manager in Sacramento." Then Loren pushed the "End Call" button.

Loren continued talking to his assistant. "Bozo Hobbs better win big on the golf course, 'cause tomorrow he's gonna find himself without a job at Purity. Now, contact Simpson at our Philly plant. Tell him we'll be touching down in fifteen minutes and to meet me outside on the pad."

When Loren stepped out of the helo to meet Simpson, no words were exchanged; they just hopped in a golf cart and headed off into the massive complex. The Philadelphia Purity plant encompassed fourteen acres just north of the international airport, next to the Old Fort Mifflin Museum on the Delaware River. Because of the drought, the river had become too low to draw from, so the Philadelphia plant was converted in 2033 from a grey water facility to a black water filtration facility.

Riding through the large rooms into more large rooms, all filled with huge pipes, Simpson finally said, "I'm going to take you straight to the problem, sir."

"Good," said Loren.

Loren's cell rang. "This is Loren."

"Dr. Loren, it's Matt Damon calling you back."

"Damon, you're my PR man, aren't you?"

"Of course, sir."

"Well, I expect you to be available 24/7. Didn't I explain that when we hired you?"

"Understood, sir. I know why you're calling. I'm all over it. We've been playing it down with all the Sacramento media. Plus—"

"No, no, I want you to focus on Philadelphia first. Forget Sacramento for now. Got it? Now, send me a one-page brief on your plans, and be damn sure to include how your lobbyists are going to handle the Senators. I want it in two hours."

Loren and Simpson's cart pulled to a stop, where they met three other men, all suited in Bio-Haz gear with gas masks. Loren and Simpson suited up and all five walked through a steel door into a giant room with thousands of large pipes and holding containers. A team of men in Bio-Haz suits were already working on a large coupler between two pipes.

"Sir, this is the problem," Simpson said, as they walked up to the man working on the coupler. "The gasket in the exchange filter has deteriorated. You can see it leaking now. X-rays also show a deteriorated membrane inside. A new one is being shipped in two days, as we speak."

"Damn," said Loren, "is that why it'll take two days? Where's it coming from?"

"Our machinery plant in Germany, sir."

"I'll call Germany. We'll have it here tomorrow."

The two men were thrown backwards from their platform ladders towards Loren and the others. Sewage sludge was spewing out of the damaged coupler and splattering everyone. The noise made it almost impossible to hear each other.

Loren turned his back, but it was too late. He and the others were covered like dark chocolate, except this was raw sewage.

They all gathered as far away as possible from the pipes.

"Why didn't you shut it down?" Loren screamed.

"We can't stop, sir. All systems will grind to a halt. Sludge coagulates unless it keeps moving. Coagulation will permeate the entire system, lock it up, and we would be down for months. We have to run a by-pass around the coupler and repair it."

"Why wasn't I informed about this?"

"I don't understand, sir. Are you asking about this problem, or the fact that it could happen?"

"About the freakin' fact that it happened! And, yes, could happen again."

"Sir, it's in the manual, but we've never seen it this bad."

Loren said, "How the hell could you have missed it?"

Loren's next questions were swift and designed to get quick answers so he could get the hell outta there.

"The new membrane," said Loren, "who's the shipping company?"

"It's coming by air. FedEx," Simpson said.

"Good. When did you post this problem on our engineering chat site?"

"This morning, sir, 0-five-hundred."

By this time, they were all covered, head to toe, in dark sludge.

"I thought so. Now I'm getting the hell out of here. Simpson, you're staying here until this damn problem is fixed."

Loren trotted as fast as he could to the clean room. He entered a washing station, where he was hosed down and chemically sprayed. After removing his suit, he noticed a four-inch tear under his left forearm. "Crap, what's this?" he asked the attendant. "What the damn hell? Did you give me a suit with a tear in it? What should I do now?"

"Sir, you'll be okay, you really have to ingest bacteria to get it in your system." Loren knew the attendant was lying to save his own hide. From then on, he wondered if the bacteria would enter his body.

Loren went through the shower area where he scrubbed his left forearm until it turned bloody red and stung. He felt satisfied.

– 14 –

The Wedding

TRADITIONAL HINDU WEDDING CEREMONIES can last for days. Or, circumstances might call for an abbreviated tying of the knot. For Jaiman and Anashka this was such an occasion. Long or short, Hindu weddings take place outside, on the earth, under a canopy called a mandap. Under the large mandap, in the center, a fire is built, called a marhwa. The fire is the centerpiece of the entire ritual. It is the *fire* that invokes the Aryan deities to cement the bonds of alliance. A Hindu wedding is essentially a fire-sacrifice. Mats can be laid down in the mandap, but no shoes or sandals are allowed. If a priest is not available, the Circle Leader can perform the ceremony. Manu's role was to act as Master of Ceremony, reading Sanskrit and quietly instructing the bride and groom through the proceedings.

One feature of the bride's wardrobe is the use of henna to decorate her hands and feet. This is done by her soon-to-be in-laws. It's said that you can tell how well a new bride is treated by her in-laws by the length of time it takes the henna to wear off. Anashka had received her hands and feet henna treatment the day before. It was obvious her in-laws approved very much of Anashka, as the intricate henna paintings on her were dark and rich, covering her entire hands and feet.

Jaiman and Rashman were still working on the perimeter reinforcements.

"You can only dream of how beautiful she must look," Rashman said.

"I can only dream of what the Bhutanese might do to us tomorrow. Or even tonight," Jaiman said.

"Keep your mind on her. Tomorrow will take care of itself. Go! Get yourself ready. You're getting married in two hours. We will take care of the rest."

"There's more time," Jaiman said. "I'll help you finish this section, then I will go."

Jaiman was toiling in the hot sun, anxious to get the perimeter protection finished.

One of the bride's maids approached Jaiman, "It's time," she said. "Your bride is waiting."

Rashman gave Jaiman a fun-loving push. "Go, prepare yourself, before she changes her mind."

Jaiman left, on his way to quickly bathe and dress in the traditional Hindu groomsman's kafni. Glancing over his shoulder at Rashman, Jaiman said, "Never, she loves me too much." They both laughed until Jaiman disappeared into the bevy of six beautiful women, who escorted him to the bathing hut.

All perimeter work stopped. The entire village migrated to the mandap to secure their places. An hour later, Jaiman and Anashka entered the center of the mandap. She was adorned in the red and white sari.

The groom wore the Kafni—an ensemble for men consisting of the Kediyu (shirt) and the Kafni (pants) used typically in the performance of the Garba and Dandiya dances.

Manu held the Sanskrit and spoke: "We have come together to wed Anashka and Jaiman. Today they build together the foundation of their marriage upon the earth, in the presence of the sacred fire and the radiant sun."

After the exchange of garlands, Manu continued: "A circle in the symbol of the sun and the earth and the universe. It is a symbol of holiness and of perfection and of peace. In these rings it is the symbol of unity, in which your lives are now joined in one unbroken circle, in which, wherever you go, you will always return to one another and to your togetherness."

Family members wept at the beauty of this saying. After the rings were exchanged, Jaiman and Anashka walked around the fire four times. Tradition dictates that the first one to sit down after the four walk-arounds will be the boss of the family. Jaiman was far too quick for Anashka. The crowd cheered at Jaiman's good fortune. Anna, Lisa, and Lynn were in the crowd, witnessing their tenth wedding in Kupup.

Boom! A blast in the distance sent everyone to the ground in panic. The wedding party began to scatter. A villager, Fallil, ran in to announce that it was an accidental explosion—a false alarm.

"Honorable Manu: It was an accident. From us, not the Bhutanese. We accidentally—"

"Go back to your position, Fallil," an angry Manu said.

Everyone re-assembled.

Everybody was happy, except for Anna. Brooks and Rinku's departure weighed heavy on her.

Bride and Groom seemed embarrassed by all the attention. Musicians played the traditional Indian wedding songs. It was a joyful occasion. The crowd danced and the sun set as Jaiman and his bride, now inside the "honeymoon tent," drank from the traditional wedding cups and spoke of children and their future together. Jaiman removed his garland, then hers and placed them on a bedside table. She smiled. They drank the last of the wine—a gift from her father.

The sitars and tabla drums from outside were music to their ears.

– 15 –

Dinner Alone

AFTER MY MEETING AND DRINKS with Gerritsen and Beasley, I was exhausted. I dropped by my favorite neighborhood restaurant, Sushi Sue. It was late so it wasn't crowded. A couple sat in the back booth, intent on studying each other. Four others were chatty and drinking cocktails. I sat at a four-seat table and ordered a glass of Lorton Pinot Gris, then turned my attention to the menu.

When the waiter returned and poured the wine, I took a closer look at him. *Damn. It's the same waiter Steve Muller and I had at Le Café Descartes that night. Damn. How could he also be working this restaurant? Could he be following me? No, wait. He's not following me, he's ahead of me. How? How could he know I'd be here?* Then I remembered mentioning my plans for sushi tonight to one of the Senators after the meeting. *Which one? C'mon, think. Which Senator asked me out for a drink? The same one I mentioned eating sushi in my neighborhood to. Which one was it? Think!*

My concentration on any Senator was broken when I saw a man walk in—truly ugly, greasy hair, pockmarked face, thick eyebrows, beefy nose. He was alone. *My God, am I imagining things? He looks like the mafia.*

I was giving serious consideration to leaving when the waiter returned to take my order. I looked over the menu then ordered tuna sashimi with Napa rolls. "Oh, by the way, did you used to work at Le Café Descartes, like just a few weeks ago?"

He never looked up from his order pad. "No, ma'am, not sure I know the restaurant." He thanked me and left. *Ridiculous, everybody's heard of Le Café Descartes.* I peeked at the other man, four tables over. He was reading on his tablet, not the menu. Maybe I'm just paranoid; maybe he's a writer. They always seem to eat alone and have something to read. The man ordered, and was served soon after me, but I wanted to see what he would do if I stopped eating, paid, and left. My waiter was too slow to bring my check so I calculated what it should be. I left fifty cash on the table, stood, and started to the front door, where I glanced over my shoulder and saw the man, his dinner half eaten, nodding to the waiter.

Instincts bubbled into a volcanic adrenaline rush as I moved with purpose to the street. Grandfather's house was only three blocks away. No time now for ride share. I ran—remembering that I'd have to run through the park near home. I made it and was grateful. To whom, I don't know. Just the cosmos.

At the front gates, I called Alfred from my phone.

I went straight to my room. Everybody else must have been asleep. Dog tired, it was time to close the curtains. Curious and still a little spooked, I looked out the front window towards the fountain. Nothing unusual. My heart rate was beginning to level out. Standing at the window, I was nonchalant about closing the curtains, thinking only about a good night's sleep. The shade's little pull knob fell to the floor, so I bent down to retrieve it. Above my head was the slightest sound of breaking glass, so soft that the thought of danger still didn't register. Looking up, I saw a small hole in the glass, and quickly dropped to the floor. Another glass break! Panic-stricken, I wondered how much more adrenaline my body could produce.

I crawled to the cell in my purse and dialed 911. "I've been shot! My address is 5408 Highland Drive, Chevy Chase."

"Are you hurt?"

"No, but I've been shot at! By a man in my front yard!"

"Stay calm, lady. What is your name and address?"

"I told you, 5408 Highland Drive."

"Got it. Your name, ma'am?"

I gave her the info and she told me to remain calm, stay on the floor, a squad car was on the way.

"Please. Hurry."

I went across the upstairs hall to wake Grandfather. Upon hearing the story, he was just as upset. We waited in his room for the police.

After they arrived, confirmed the bullet hole in the window, and found the spent bullet in the wall, the sergeant told us, "Yep, high-powered sniper rifle. Looks like a professional hit. Would have blown your—well, never mind. Thank goodness you're okay ma'am."

The police stayed and asked an hour of questions—all the who, what, when, and where I had been doing that night, last week, practically my life history. Finally, at midnight, they said they would station a policeman outside, "just in case."

I went to bed thinking it couldn't have been Muller's doing. He needs me in India. It has to be Loren. *Who else would want me dead? A senator? Maybe. Somebody sent that creep at Sushi Sues.*

Finally, I fell asleep, not knowing the time.

– 16 –

MULLER AND HIS MEN

BERNARD LOREN, BEN ADAMS, HOUSE legal counsel, and Mike Tagg, EVP of Operations, were ushered into Muller's outer office, an employee museum full of exotic animals.

Muller had not yet arrived through his private entrance.

The maze of rare animals was a taxidermist's dream of African, Indonesian, and South American specimens, full bodied and in motion—all "Big Tom" Muller's hunting trophies.

The center piece was a majestic lion in full attack mode stationed on a pedestal turning round and round, the hum of an electric motor the only sound.

There were at least a dozen other animals—leopards, hyenas, tigers, all inherited from his father. Having seen these a hundred times before, Loren, Adams, and Tagg moved quickly to Muller's private office.

Two chairs were stationed in front of Muller's aquarium desk and a sofa off to the side.

Loren always took the sofa, leaving Adams and Tagg with chairs in front of Muller and his desk. They'd all seen the piranha water show, so they knew it wouldn't be on display today.

To Muller's left was a giant electronic LED board, showing Purity facilities all over the world. Little red blinking lights showed where land and water tankers were located, their destination, and ETA.

Muller marched in, mouth clinched, not smiling. He twirled his chair around and said, "Would someone explain to me how the hell we're going to prevent another shutdown like this?"

"Steve, we've ordered and stocked back-up gaskets and membranes for every plant," said Tagg. "The new facilities coming on-line next month will have back-ups. And the MIT team swears to me they're 90 percent finished with the new membrane technology. They claim it has a ten-year life span no matter how much we push through it."

Loren chimed in, "Ben, American stole our first patent. What are you doing to stop them from getting their hands on this one?"

Muller didn't wait for Adams to answer. "No, wait, I'll tell you what we'll do. We'll shoot them," Muller said.

All three men shot Muller a concerned, quizzical look.

"And knowing we're gonna need a trigger man," Muller looked over at Tagg. "Tagg, you're it."

Tagg's eyebrows shot straight up as his eyes popped open and revealed more white than any of them had ever seen. He was red as a beet and shocked.

They all laughed.

"Maybe not a bad idea," said Loren.

More laughter.

"Seriously, Steve," said Tagg, "you could've given me a heart attack. Trust me, the MIT team is under lock and key—at our warehouse in Tampa. Our security has tapped the cells of every MIT team member. Plus, I've got a woman, a scientist, on the inside at American. We'll know when and if they go into R&D with a new membrane."

Adams jumped in, "And Steve, I've already filed for a provisional and I'll file for the full patent once we get the artwork completed. That's the best we can do for now."

"Okay, what's Sacramento look like?" Muller asked.

"Temporary fix on the system for now," said Loren, "but we should be at full capacity in one week. We've switched back to grey water until the problem is solved. I was working with Hobbs' number two man in Sacramento. But like you thought, he's not our guy. We're searching for a replacement."

Pointing to the LED wall Loren said, "I want to talk about distribution for a minute. We're spending way too much on shipping. Fuel is at an all-time high and going higher. We have 462 tankers traveling the globe. I've run the numbers, and if we build five new plants, just five, all overseas—I recommend Cape Town, Rio, Riyadh, Kiev, and Melbourne—we could save a minimum of twenty billion dollars over the next three years."

"Let's do it. Get it underway," Muller said. It was nothing for Muller to make billion-dollar decisions that fast. He'd been doing it for twenty years.

"But those are all wealthy cities," Adams said. "Why can't we locate a few plants in at least a couple of desperate cities?"

Tagg jumped in, "Because we need new plants located where our distribution is weakest. We can compete on price if we get our distribution costs down."

"So basically, 'screw the poor areas?'" a concerned Adams asked Tagg.

Muller cast an evil eye toward Adams.

"Ben let's not slow this meeting down with a moral discussion of who gets what water. Don't worry, we'll have an affordable price for those spots."

"Yeah, with poor quality water," Adams complained, looking over at Tagg.

Muller's lips tightened into a razor thin line. "Careful, Ben. You know our brand promise: *Purest Body Water in the World.* I'm proud of that."

"What's the upshot of the Philly situation?" Adams asked. "Our stock dropped, but not as much as you thought, right Steve?"

"No, thankfully," said Muller, "we've solved that problem. The PR firm came through. Berney, nice job on your part."

Loren acknowledged with a head nod.

Muller aimed a laser pointer on the Himalayas, "But that, gentlemen . . . that is our greatest potential problem. The Water Sustainability Committee, then Congress, needs to know—once and for all—that the notion of oceanic cloud seeding is simply not a viable possibility. Black water filtration must remain the primary source for body water. We will not survive a vote for nationalization unless we can prove that seeding clouds in the ocean is a fairy tale.

Loren said, "Six failed experiments should do it. We just need to get Schwarzkopf, and not the girl, over there so we can make sure every experiment is a bust."

Muller knew how he was going to get Schwarzkopf to the facility, but he wasn't ready to reveal it to all three men.

"Gentlemen, that should do it for now. Berney, stay for a minute, would you?"

Adams and Tagg left, muttering something about the "Tagg you're it" joke.

Muller's assistant brought him a tux shirt, while Loren watched Muller struggle with the studs. "Berney, since Bennett's not got the oranges to get Schwarzkopf to work on his theories at our India facility, I have an even better idea."

"Don't tell me it involves his granddaughter, Grayson Fields."

"Berney, I don't care how you found out, but trust me, it's the only way. She'll come first, and he's not gonna let his baby granddaughter hang out with us for very long, do you think? He'll show up."

"I don't like it, Steve."

"Why? What could go wrong?" Muller's assistant is now pushing the last stud through his shirt, as he squeezes her derriere and says, "Thanks, lovely. You're my girl, right?" She smiles and leaves, cleavage and all.

Muller then turns back to Loren. "Look, even if he doesn't show, all we need is the algorithm, which we're gonna make sure ends up in our hands."

"She could screw everything up."

"Screw it up? How?"

"She could figure out how to do it."

"The only thing she's going to screw is me. Look, Berney, c'mon, you really think she could solve the cloud-seeding problem? You've got to be kidding. Hell, even she doesn't believe in it. She can't solve it. And neither can he. She told me. It doesn't work. Which, as you recall, Bernard, is the same as you told me three years ago. Like you said earlier, we just need six failed experiments before the vote comes to the floor and we'll be done with this gauntlet. Anyway, look, the deal you put together with Kenya the other day will add five percent—overnight my friend—to our revenues." Muller grabbed Loren by the shoulder and shook him with glee. "What you did is what counts the most. Let me take care of Washington, you take care of operations."

"Steve, once again, that's not exactly how we discussed our roles here," Loren said.

"Berney, Berney, listen to me, we've been at this together, what, ten years now? It's time you were promoted to CEO. I need to spend more time on the hill. We'll talk about details later, okay? I'm meeting Grayson at the front steps of the Kennedy Center in thirty minutes. I'll see you there, right? Right?"

"Of course," Loren finally said. "Home to change first. Steve, thank you for the promotion. But please, trust me on this about her. It's risky to involve her."

On the way out, Muller laughingly shrugged him off and pointed to the Himalayas on the LED wall. "Berney, don't worry about Grayson."

He smiled and shut the door behind him.

Disgusted, Loren slung his tablet at the sofa.

– 17 –

TIME's Person of the Year

Early evening, front steps of the Kennedy Center

My limo door opened. Muller was standing on the sidewalk, waiting. I was in my little black dress, low cut, with pearls, professional make-up, and a new hairdo. I knew I looked good, and slightly older.

He reached out to take my hand. "You look beautiful."

"You look like *TIME*'s 'Person of the Year.' *Why do men look so good in a tux?*

We walked up the red carpet into the reception, my hand on his arm into a cast of hundreds—Senators, celebrities, even The President and First Lady.

People were sipping champagne, mixing and mingling. We weaved our way through the crowd, Muller accepting congratulations at every turn. In between these praises he whispered in my ear.

"Any chance you missed me?" he asked sweetly.

"Any chance you missed *me*?" I whispered in his ear, playing the role.

"Fair enough," he said. "And the answer is, yes. I've missed you. Somewhat."

So, after one dinner, you've missed me? You lying bastard.

"Since we're being so honest with each other—" I started to say but was interrupted by yet another patron paying homage to the world's richest man.

"Steve!"

"Hello, Ray. How are you?" asked Muller, casting a toothy business smile.

"I'm doing better than I deserve," said Senator Gill. He turned to me with a genuine smile. "Good to see you again, Dr. Fields."

We shook hands. I nodded. "Good to see you." And I meant it.

Muller acted surprised. "Well, I'm glad you two know each other. What's up, Ray? Wait, don't tell me, more water problems in Arizona?"

"Steve, this may not be the place to talk, but I'm deeply concerned about my constituents in Arizona. I've received too many emails to ignore this—this situation." He shook his head, his mouth down in disgust.

"And your question is—?" Muller asked.

"Mind if we step over here, (pointing) for a little privacy?" Gill asked.

I went with them to a corner nearby, partially covered by a giant ficus tree. Fake, of course.

Gill took a white handkerchief from his coat pocket and wiped a moist forehead. "Steve, families are complaining. Hell, they're at my doorstep. They're angry and they're making accusations connecting your water to dozens of severe health cases. Steve, people are getting really sick. I need to know what can be done about it before I call for a Senate investigation. You realize what a pandemic alert from the CDC would do to all the water companies, don't you? Aren't you still president of the association that represents filtration companies?"

"Yes. But gracious, Ray, don't you think it's a little harsh and even hasty to blame Purity? Look, let me set up a meeting through your staff chief. I'll have some answers for you. Whatever the problem is we'll get it fixed. Will that work?"

"That'll work fine," said a polite Gill. "Oh, and, Steve, congratulations on tonight. It's a big honor."

"Ray, please keep in mind, the problem could be American h2O's, not ours; you're aware of that, right?"

"We'll see," Gill said and gave a polite goodbye. We continued our casual stroll through the crowd.

I asked Muller, "What's wrong with your Arizona water?"

"Don't have a clue," he said. "But I'm going to find out. Now, what were you about to tell me? Something about being honest with each other."

"Yes, I'd like to have some idea of your plans for me."

His sexy smile answered my question. "No," I said. "Not like that. About what you want me to do for *Purity*, not for you." I hoped my cat-like smile would keep him interested but not promised.

"I thought I'd already told you," he said. "I want you to come to our facility in the Himalayas and help us solve the ocean cloud-seeding problem."

Well, that settled it for me. I don't need to ask for the facility assignment. He offered it to me.

"Of course," I said, not wanting to look excited. "But you know I don't believe oceanic cloud seeding is possible. I've already told you that."

"Just try, that's all we ask."

Here we are, two liars, both knowing the other is lying. But I suppose it's the game you play if you're going to survive in Washington. "And Grayson, there's something else you should know. Dr. Loren doesn't trust you. So, be a good girl and get on his good side, would you?"

I grabbed his elbow and leaned into his right ear. "If you ever refer to me again as your 'good girl' you'll never see me again. And why are you telling me this about Loren? Do you think I give a crap about pleasing him?"

I think I surprised him with my sudden bravado. He leaned back, as if pretending to be frightened of me. "Nooo. I just want you to be aware of what you're getting yourself into." He seemed sincerely remorseful. Still, I felt like taking a shower as soon as possible to wash his filthy lies off me. Of course, *my* lies were justified. I began to realize how manipulative Steve Muller, "Poseidon of the Waters," could be. And how counter manipulative I could be. It gave me a rush of adrenaline. *I think I could make a pretty good spy.*

Maybe there's a tiny part of him that wants to believe me—even though I'm going to bust my butt to make oceanic cloud seeding work, instead of trying to prove it doesn't. I already realized how tricky that was going to be. And dangerous.

I asked, "Why would Dr. Loren, one of the most prestigious hydrology scientists be concerned with a peon like me?"

"He thinks you can figure it out. He knows you've been working on it with your grandfather for several years. He thinks you can actually make oceanic cloud seeding work."

"He's wrong. It'll never work."

"Let's just make sure."

There it was—the drive in his desire to have me on sight at his facility. He had never come straight out and admitted that he wanted all the experiments to fail. I'm not sure he caught his own inadvertent faux pas.

'Course, I knew this was his plan all along. I'd just been caught up in the cloak and dagger and forgotten my mission. Surely, I'm back on track now. He and Loren will try to sabotage whatever I do. "Yes, thank you," I said, taking a glass of champagne from a roving cocktail waiter.

Five minutes later, Bernard Loren himself stepped in and interrupted, "Congratulations again, Steve, it's going to be a big night. Oh, sorry, did I interrupt something?"

"Not at all," Muller said. "How was your trip to Philly?"

"A real bitch. But we'll be back to normal soon."

Loren turned to me. "And you, Miss Fields? How are you?"

"I'm fine, thank you."

Some people are just naturally rude. They make eye contact for a millisecond, pretending to be interested in you, but after that they're looking elsewhere over your shoulder to find someone more important to impress. Bernard Loren is the poster child for that sort of boorishness.

Low and behold, an attractive young woman walked up. "Aren't you Dr. Loren?" she asked.

"Why yes, I am. Do I know you?"

"Not yet. I'm a reporter for The Capitol Hill. May I ask you a few questions?"

"Fancy that," he said, "What excellent timing. You saved me a phone call to your publisher. I was going to have a word with him about some gross inaccuracies of late in your publication. Step over here and I'll help you salvage The Capitol Hill's reputation."

As they walked away, I muttered, "It's *Doctor* Fields to you, Mister Loren."

Muller heard me and laughed. "I see you two will get along fine."

"Your sarcasm is not appreciated."

Several more Congressmen, Senators, and celebrities approached Muller as if they were entering the presence of their master. It was his stockpile of money they were after.

Muller whispered in my ear again, "I'd like for you to get up to speed on our simulator in the mountains. Are you prepared to go there next week?"

"Steve, I'm thrilled, but, no, not next week. I need a few weeks. I need to sever my ties with a stubborn old man."

(Again, I hated to lie, but lying is acceptable when lives are at stake, like if I were a spy under interrogation and knew the location of every spy in my network I would definitely lie before telling the interrogator their locations.)

Besides, I needed time to learn about the design straight from NOAA in their D.C. offices before I landed unprepared at the Himalaya site.

"Give me just a few more weeks, and I'll be ready."

Just then, the lights in the great hall dimmed, brightened, and dimmed again, signifying that the dinner and presentation was about to begin.

I don't remember much about the dinner or Steve's acceptance speech. All I could think about was my upcoming meeting with Pete Gerritsen and his staff at NOAA.

NOAA is practically the only agency knee deep in figuring out this cloud-seeding problem. Why are scientists still looking for life elsewhere while we're in need of life support, right here? And politicians continue to line their pockets with pork contracts, celebrities are looking for new stages to preen their political hair. Why we put celebrities on a pedestal and let politicians steel us blind with their pork is beyond me.

7:25 a.m. I called Pete Gerritsen, head of NOAA. "I'm ready, Mr. Gerritsen. When do you want me there?"

"Day after tomorrow, 10 a.m. I'll have the team ready for you. And Grayson, if you ever call me Mister Gerritsen again, I'll have your name removed from my contact file." He laughed. "Is that clear?"

"Yes, Sir—Grayson Fields, reporting for duty."

I was eager to go over the design of the giant simulator.

– 18 –

Bhutanese Warriors

Rinku ran hard, ducking, jumping, dodging limbs, logs, dry creek beds, slipping past anything his skinny body encountered. Exhausted, dirty, sweating profusely, he wouldn't let up.

He had forgotten that a wedding ceremony was likely underway back in his home village. As the sun was setting, he kept moving through the dead forest. Sensing he was near his village, he spotted something. Something wrong. The moonlight revealed men in masks, crouched behind trees.

They're on the wrong side of the trees to be Kupup lookouts, he thought. He went to a low position, lower, lower, crawling, listening. He heard the language of Dzongkha. Sweat continued to run down his face and body. He couldn't control his breathing.

He realized these were the Bhutanese warriors from the village they were supposed to find. Some wore wooden face masks depicting demons, death heads, and animals.

He looked down into the village, fifty yards away, to see the sacred fire burning beneath the mandap.

In almost complete darkness now, most in the village had retired to their huts, but some were still dancing and singing with the musicians. Nothing happened for a few minutes.

Then all hell erupted, as Rinku watched in horror. Hundreds of Bhutanese warriors came from behind the trees and raced down to the village. His village. He heard death screams before they reached the village. Near him, two Bhutanese fell into the Cobra pits. He heard their screams.

Through the thick dry timber, Rinku saw his own villagers toss torches into the trenches. As the fire was beginning to blaze, hundreds of Bhutanese ran into and out of the flaming trenches. Some caught fire, but most made it through before the trenches could develop a decent wall of fire. Dynamite blasts blew up several dozen Bhutanese on Rinku's side of the village. He heard blasts coming from the other side.

Rinku crawled to a closer, better vantage point. He was ready to run down and help. There was chaos. Spears being thrown, swords being thrust into bodies. Rinku realized he was no match for these warriors. He stayed put, feeling fear, and guilt.

After five minutes, the Kupup warriors were on their knees around campfires. Bhutanese warriors went throughout the village gathering up the women and children. Rinku knew they would become slaves.

He saw four Bhutanese warriors drag a nude Jaiman and Anashka out into the open and under the mandap, beside the sacred fire. He couldn't believe his eyes. He crawled closer, waiting for the right opportunity to go down and save the village. But the Bhutanese wasted no time in running their spears through Jaiman and Anashka.

It was dark, but the campfires lit enough faces for Rinku to recognize Rashman. He was arguing and screaming at the Bhutanese chief, who said something to Rashman, making him lie down on the ground, where he sobbed. Rashman was not tied, cuffed, or harassed in any way. He just sobbed. Rinku could see this. At first, he had no idea why Rashman was not murdered along with the others.

He reflected on the conversation between Jaiman and Rashman. Rinku had been close enough to hear bits and pieces of Rashman telling Jaiman about being in love, but never told Jaiman her name. *Was she from the Bhutanese village where he had gone to spy? Was his life about to be spared for supplying them information about Kupup?*

Sick to his stomach and sick in spirit, Rinku puked in his hand, covering his mouth to avoid any noise. His eyes were bloodshot red and wet with tears. Rashman had been one of his mentors. Rinku had wanted to fight like Rashman. Not anymore.

Rinku watched as Manu was brought under the mandap, next to the dead bodies. Bhutanese warriors surrounded him, dancing, waving their spears and swords. Their torches caught the mandap on fire and was quickly torn down, still in flames. He saw the church tent on fire. The chanting they made was the worst for Rinku. He had never heard such savage sounds.

They pushed Manu down to his knees. Looking up to the full moon night sky directly above him, the Gedu chief shouted something. Rinku couldn't make it out. Then he heard the leader say: "Our mighty god, we offer you this sacrifice as a pledge of our faith to you in return for the famine you have spared us from. You have given us water, oh god, that we may survive and multiply again."

All the warriors screamed to heaven. The chief nodded some sort of approval to his men. They all moved in on Manu and each acted as if they were going to thrust a spear into him; but abruptly cast their spears into the dirt, each beside his right leg.

Rinku was about to move to another position to somehow get in and find Anna and the girls. Before he could move, he saw them bring a woman and two girls out into the opening and make them stand side by side. Rinku wanted to scream. He was certain this was Anna and the girls. They were just too far away for him to know for sure. He wanted desperately to run down into the village, cut the Bhutanese's throats and run away with Anna and the girls. *Oh, my God, it is them*, he thought.

The invading chief came over. He gave an instruction to his warriors. They immediately ran spears through all three of them, Anna, Lynn, and Lisa.

The Bhutanese were not normally warriors with a blood vengeance. They were survivors in a new, Darwinian world, in search of life-giving water.

Reverend Brooks Turnage was still in the village of horrific water-disease deaths, on his knees, screaming at the heavens, "Rinku, where are you? AnnAAA!"

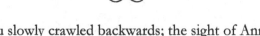

Rinku slowly crawled backwards; the sight of Anna, Lynn, and Lisa, made him heave. He took a last look at what remained of Kupup, and crept away, sobbing.

After a day-and-a-half of running from Kupup back to Chhukha, Rinku looked half dead, barely able to walk into Brooks' arms. They hugged and cried. Brooks held Rinku out at arms-length to look closely into his eyes. He saw what Rinku was unable to tell him. Brooks instinctively knew he would never see his wife and little girls again. He pulled Rinku to his chest. Tears streamed as they each stared off into the distance.

"We're going home," Brooks said, tears flowing, "home to London."

– 19 –

LONDON

BROOKS AND RINKU SAT IN a musty old office of the London All Saints Church on Manchester. Reverend Ellis Wright, practicing psychologist and Chairman of Missions Outreach for the church, was doing his best to help Brooks with his grief. The loss of Brooks' wife and two girls had taken a toll. As Brooks explained to Wright, there was no rhyme or reason why God would take his family from him after his laboring for so long in such a remote part of the world.

"I did. I signed up for wherever God needed me," Brooks said. "You sent me to some God-forsaken village in—in bloody nowhere!" Brooks continued his cursing, now manifested in full bloom.

Rinku was taken aback by Brooks' coarse language. Wright, not so much. He had heard much worse grief regurgitations in his fifteen years as head of missions.

"Are you offended by my language?" Brooks asked Wright.

"Not at all. You wouldn't believe what I've heard from missionaries who've lost families. Listen, Brooks, I want you to tell me everything you're feeling, right now, right here."

"Why?" asked Brooks.

"Because it's the best thing you can do."

"You're telling me the best thing I can do is be angry as hell with God?"

"Yes, he wants you to spill your bloody guts out to him. Believe me. You cannot offend him. You cannot. Not if you're one of his children. He wants you to be honest with him."

"Then let's get Rinku out of here, because he shouldn't hear what I have to say if that's what you want."

"It's not what I want, it's what God wants." Turning to Rinku, Wright asked, "Rinku, do you wish to stay, or go upstairs?"

"Sir, I would like to stay. Reverend Turnage—he is like my father, only he is not my father. He's the only family I have."

Staring at Rinku, Brooks tried hard but couldn't stop the tears from welling up in his eyes. "Brooks," said Wright, "you know we have a special adoption procedure here at All Saints."

Brooks sat erect; his expression turned to anger. "What? You want me to give Rinku up? There's no—"

"No," Wright interrupted. "I mean you can officially adopt Rinku as your own son."

Brooks' expression changed from absurdity to hope. He turned to Rinku. "Rinku, would you like to be adopted? Would you have me as your father, an old cuss like me?"

"What's a cuss?" Rinku asked, smiling. "I think I like it, whatever it is."

They all laughed.

"You'll arrange it?" asked Brooks, turning to Wright.

"Yes, most definitely."

"Good. But Ellis, I must ask you ... you said God could not be offended by my honest—and I mean *honest*—feelings. How is that possible? How is it possible for God to not be offended by my bloody awful feelings about him?"

"Because, if you are truly one of his children, you can always open your heart to him. However, if I were you, I wouldn't go around cursing him, if that's what you're asking."

Wright continued, "Brooks, you know the story of Job. At first, he doubted God's wisdom. Now you'd think God would throw him under the bus for something like that. God reprimanded him all right; but in the end, God gave Job back all that Satan had taken away, plus twice as much as he had before! It's a story of trusting God.

"Brooks, look, there's a book I want you to read." Wright went to a shelf on the wall and pulled one out. "It's an old book by a theologian named R. C. Sproul, *Surprised by Suffering.* Would you promise me you will read this book, and meet with me to talk about it?"

Brooks studied the cover, thumbed some pages, and said, "Sure."

"Brooks, I want you to talk about Kupup, the village—what it was like. The people. Their circumstances. Paint me a picture. It's my job, you know, to gather this information."

"No."

"Brooks, you need to talk about it," Wright said. "Trust me, it's the best thing you can do."

"Best thing I can do?" asked Brooks.

"Yes, for your healing, and for your recovery, and for your next mission."

"There won't be a next mission. What don't you understand happened back there? I lost my wife! I lost my two precious girls. And I had no converts. What more does God want from me?"

A pregnant pause between them. Wright had never been stumped before and was determined not to be.

"Okay. Tell you what," Wright said with a stern voice, "when would be a good time for the adoption process?"

"Tomorrow?" asked Brooks.

Wright smiled. "Brooks, you *are* a man of action, aren't you?"

Brooks didn't respond.

"Tell you what, let's make a deal. I'll pull a miracle to have the adoption take place tomorrow if you'll agree to come back the day after and fill me in on Kupup. You realize it's part of my responsibility here as missions chairman to record the debriefings of our missionaries? It's to help other missionaries going into the field."

"Okay," Brooks said.

"Good. I'll call you in a few hours with details of the adoption. Brooks, I'll be pulling some major bloody strings here, so I hope you'll live up to your side of our bargain."

Brooks nodded, as he stood and left with Rinku.

Two days later, Wright asked Brooks, "So how was the adoption?"

"Fine," Brooks said, still not much for conversation.

"Well, congratulations, I pronounce you 'father and son'."

Brooks and Rinku looked at each other with toothy grins.

"Now, I believe we're here together to talk about Kupup, is that right?"

"I suppose," said Brooks.

"Okay, well, then, why don't we start with the village, the people. Tell me about them."

"There's not much to say," Brooks said. "There were about two hundred people. Very agrarian if you can call growing crops without water, 'agrarian.' Well, actually there was *some* water to be thankful for. We had a well, thank God. I helped them dig another one. It wasn't much, but it helped. Sometimes the women still had to fetch water from a spring about three miles away. Our fifteen-liter containers should be reduced to ten," Brooks said. "Fifteen liters is back-breaking work for the women. Would you make a note of that?"

"I will, indeed," said Wright, tapping away on his notepad.

"I set up a health clinic, like missionary school taught me. Anna and the girls—they—" (Brooks dropped his head onto his chest and tried his best not to break down. He rubbed tears away and stiffened. Wright gave him time to collect himself.)

"We all pitched in, and the medicines were quite useful. Typical problems, you know? Toothaches, stomach, some broken bones, the usual. Had no real serious sicknesses. We were fortunate in that respect."

Wishing to lighten the mood, Rinku broke in, "Father, tell Reverend Wright about the man bitten by the cobra?"

"Cobra?" Wright asked, surprised.

"Oh yes, that man," Brooks remembered. "We thought he was going to die. I didn't realize cobras came in so many varieties. This nasty bugger was green."

"I'll put 'cobra training' in the curriculum," Wright added. "What happened to the chap?"

Rinku started to giggle. "It bit him on his—on his member."

"His penis?" asked an astonished Wright.

Brooks broke in: "Well, the fool man was acting like the thing was a pet. He was playing with it in his lap. He probably wishes he were bloody dead."

Wright laughed, Brooks only smiled, while Rinku looked as serious as ever.

"Tell me more about the people," Wright said. "What about converts? No converts for the record?"

"We built one of our standard tent sanctuaries. It wasn't much, but it was good enough. I held services every Sunday and preached. We sang songs. Typically, only a few would show up—well, besides Anna and the girls, and Rinku, of course. Mostly, we cared for the sick."

"What about the converts?" Wright asked again.

"Like I said, not many believed in Jesus. I'm not sure I had any converts." Brooks hung his head. "Five years, Ellis—five years."

Wright could sense the failure on Brooks' heart. "Does that bother you?"

Long silence.

"Yes, very much," Brooks said.

"Why? Why does it bother you?" Wright asked.

"Because, well, isn't that what we're supposed to bloody well do?"

"No, not really. You remember what you learned in seminary, don't you?

"I know, I know. We sow, he reaps," Brooks said.

"Brooks, right now, while you're here, there are many souls in heaven that you don't even remember preaching to or loving on. You've been on the front line. God has surely blessed your efforts and your sacrifices. You can trust in that!"

Wright's pep talk didn't faze Brooks. "Look, I'm going to resign," Brooks announced. "Rinku and I will live in London. If there's not a position here in the church, I'll look for other work."

"I'll see what I can do but I'm afraid our budget won't allow any more hires. I know of a company that might need you though. I'll check, okay?"

"Thank you, I appreciate it."

Brooks and Rinku returned to their room for their fifth and last night stay at All Saints Mission Church.

– 20 –

LOREN'S STORM

LOREN SUMMONED HIS LIEUTENANT. "I'm scheduling another experiment for September 11. This time I want you to double the amount of silver iodide and phosphorus. We're going to create the biggest storm ever over the Bay of Bengal."

"Same procedure for notifying all flight controllers in the area?"

"No, not this time, I'll let them know. Just make sure your team is here and ready."

"Yes, sir."

– 21 –

Village Troubles

Ellis Wright was reading a book when his assistant came in with a message. The messenger told of villages in India and Bhutan that had been pillaged; he presented several satellite images. The images showed people sprawled on the ground in all directions—some running toward caves, others into the dry timber forest.

Wright gasped, "Are these our missionaries?"

"Not all of them. We don't have a post in this one here, in Bhutan. I'm not sure how to pronounce it, but I believe it's the village Brooks Turnage said he came from. I remember him saying he was in the wrong village."

"Chhukha. I believe you're right. He'll want to know about this," Wright said. "And these others, Tashagang, Gong Thong, and Samrang. Get all the information you can about each one. And get the coordinates for all four. I'm going to call a staff meeting and get teams put together to go. After that, I'll get with Brooks and let him know about this village of his."

The staff met in the large mess hall. Four satellite photographs hung on one wall. "Four villages have been pillaged," Wright explained. "We have their exact locations on satellite. I need two men to round up three missionary recovery teams; even if you need to borrow some from Mt. Vernon Church."

Two men volunteered and left.

Brooks and Rinku were bunking in a second-floor bedroom. That morning at one a.m., Wright quietly woke Brooks.

"Brooks wake up. Come with me."

They walked out into the hall.

"Brooks, I thought you should know," he whispered. "Here's a list of villages that have been pillaged. Tashigang, Gong Thong, Samrang, and this one, Chhukha. Didn't you tell me that was the wrong village you and Rinku stumbled into?"

"That can't be. That's Tahir. The Circle Leader's village. Oh, no. You're sending a recovery team over, right?"

"I'm afraid not; not to Chhukha. It's in Bhutan and we don't have any posts there." We're sending teams to the three where we have a missionary."

"Then I'll go. I'll figure out a way to get him out of there. These war lords spare the circle leaders for some reason I don't understand."

"Are you sure about going?"

"Yes."

Brooks went back to his bunk. He couldn't sleep. He woke Rinku at four and told him he was going and Rinku was to stay here. "It's too dangerous."

Rinku was upset.

"It's *Tahir!* I need to go with you, Father. I can take care of myself. You know that."

Brooks couldn't talk about it anymore, so he gathered his toiletries and said, "I'm going to take a shower. I suggest you do the same."

At six, a dozen men gathered in the mess hall. Wright divided them up by the villages he was assigning to each. He asked Brooks and Rinku to go with a young man to one corner of the room. "I need someone from our staff to accompany you."

The young man was fresh out of seminary and mission training. Brooks thought he looked far too young to be going and told Wright. Wright's response was simple: "Like I said, I need a staff person to accompany you."

"As of right now, I'm on your staff. Besides, three's too many. He would slow us down."

"Very well, then." Wright said. He turned to all the teams. "Listen up. The large map up there, on the far wall, you'll find the location of your village and a packet with all the information we could gather in just a few hours. You won't be replacing anyone when you get there. Just pray most of them are alive and you can do something to help. We'll be praying for you and all the villages."

"Is air transportation in the same place as before?" someone asked.

"Yes, it's all arranged. God speed, everyone."

– 22 –

MULLER & LOREN

THEY MET TO DISCUSS MULLER'S upcoming meeting with Senator Gill regarding his concerns that inferior water was being sold in his home state of Arizona.

Loren leaned back and told Muller the Senator must be mistaken. "No inferior water is being sold anywhere," an indignant Loren said.

Muller asked, "What about the shipment to Senegal you mentioned in our meeting the other day? How can you be certain Arizona isn't getting inferior product?"

Lowering his brow, Loren said, "Because I said so. So what? Tell Senator Gill he should check his facts. He should send his investigators to American h2O. Like I've been saying, it's probably them. Or Avanti from France. They've been 'dumping' cheap crap over here for months now."

Muller believed him.

22

MÜLLER & LOREN

THEY MET TO DISCUSS MÜLLER'S upcoming meeting with Senator Gill regarding his concerns that inferior water was being sold in his home state of Arizona.

Loren leaned back and told Müller the Senator must be mistaken. "No inferior water is being sold anywhere," an indignant Loren said.

Müller asked, "What about the shipment to Senegal you mentioned in our meeting the other day? How can you be certain Arizona isn't getting inferior product?"

Lowering his brow, Loren said, "Because I said so. So what. Tell Senator Gill he should check his facts. He should end his aversion to American H2O. Like I've been saying, it's probably Arctic. Or Aveen from France. They've been 'dumping' cheap crap over here for months now."

Müller believed him.

– 23 –

MULLER & GILL

MULLER AND SENATOR RAY GILL met for drinks at the Capitol Grill Bar.

"Scotch, neat," Muller ordered. Gill asked for a ginger ale.

"Steve, I have more proof, if you care to see it."

"Proof of what?"

"That Purity is responsible for millions of dollars' worth of damages due to tainted water."

"I don't know what to say, Ray," Muller said, shaking his head. "I came prepared to tell you that American is responsible, not Purity."

"Well, I have some photographs. Let's see if these tell the truth, shall we." Gill said.

He opened his tablet to the first photo—a tanker truck with a PURITY WORLDWIDE logo painted on its side parked at a truck stop with a Yuma, Arizona sign in the background. Another photo revealed a PURITY WORLDWIDE 40-gallon plastic container on an interstate road near an Arizona law firm billboard.

Gill clicked on other photos of tankers unloading into smaller water trucks at a fenced facility with a large sign in the background, "PURITY WORLDWIDE PHOENIX, No Trespassing."

Muller put a finger down on the tablet. "Who's to say these weren't photoshopped?"

"They were all taken in Arizona, Steve," Gill said, his mouth clenched so tight his lips disappeared. "But you're right. None of these, by themselves, prove anything. But if you look closely at the photographs, you'll find a batch number for each tanker, each barrel, and we have batch numbers for every bottle sold at several convenience stores."

Gill pulled a thick envelope from his briefcase and placed it on the bar. "Now, the proof, the connection, is all in here. Our Department of Health Sciences studied 122 cases of suspected water-borne disease. The participants in the study were asked where, and what brand of water they purchased in the last two weeks before becoming sick. They're all from Purity. We even pulled your product from grocery shelves and analyzed them. They contain a new bacterium." Gill shook his head again, "Steve, you're in trouble."

"Ray, listen, six months ago, we lost three tankers to hijackers in Utah. We thought they were water scavengers. Now I'm suspecting that American stole them and is using them to haul their own water. Ray, they'd like nothing more than to take all our Arizona business. They know how much your state buys from us."

"You're saying American stole your tankers and have been filling them with their inferior water? That has to be the biggest cock and bull phony baloney crap I've ever heard, Steve," Gill said. "Steve, do you really take me for a fool?"

"No, Ray, of course not. Look, let me take a copy of these back to my people. Let us analyze them."

"No! I've proven my case. We've spent over fifty million dollars in health care benefits to our citizens on this problem alone! I'm going to our attorney general and explain this. I'm sure he'll want to prosecute. I'm sorry, Steve, I know this is hard news, but I have to protect my state. Plus, there's an election next year."

"Yes, there *is* an election next year. And since you're playing hardball with me, do you remember what I said about Arizona being our largest customer? We employ thousands in your state. If you shut us down, not only will all those people be out of work, but there will be a large gap in water delivery to your entire state. The ripple effect from this, coupled with our PR machine, will make damn sure you're not re-elected. Ray, all I'm saying is give me four weeks to prove these are fake. I'll get it done. And we'll remain friends. You don't want me as your enemy. And I don't want to be sending tainted, diseased water to Arizona or anywhere for that matter."

Gill rubbed his cheek and stared at Muller. Okay, Steve, but you need to come back with some damn strong evidence that it's American h2O doing this and not Purity. I'll give you three weeks, not four. Agreed?"

"Sure, Ray," Muller said, his face flushed with blood, having just been on the defensive for one of the few times in his life. As Gill walked out, Muller polished off his half tumbler of scotch in one swallow. He watched Gill leave the front door and muttered under his breath, “And you, Senator Gill, have just signed your death warrant.”

Muller and Loren met in his D.C. office. Muller explained all that happened in the Gill meeting. Loren didn't seem too alarmed.

“Berney, do you see the problem here? You don't seem too concerned.”

“I'm not. I like your answer about American steeling our tankers. How’d you come up with that?”

“Never mind. It’s called ‘thinking on your feet’.” Listen Berney, we have three weeks. The first thing we need to do is back- date a police report for the theft. It needs to be on record. Can you take care of it?”

“Sure. It shouldn’t be a problem.”

“Then I'll have to think about next steps.”

“I'll be thinking, too,” Loren said.

“Meantime Berney, you're going to make sure we halt any further shipments of water to Arizona that has so much as a hint of a problem with it. Got it?”

“I'll take care of it,” Loren said. “And what about Gill?”

“Take care of that, too. You know who to call.”

“But will we still need the police theft report?”

“Yes, Gill might be gone, but someone will ask questions. We need to be neat.”

– 24 –

GRAYSON MEETS WITH NOAA

GERRITSEN HAD PREPARED FOR MY arrival. At the check point I was given the necessary credentials. No one entered the inner sanctum without a badge and lab coat emblazoned with the famous NOAA logo.

Gerritsen appeared. "Grayson, good to see you." We shook hands.

"You, too. Pete, I want to thank you for setting this up. It means a lot to me."

"Well, it's for a good cause. Before you and I go over to India, we need to know all we can about the simulator. And you're about to meet the team that designed it and is going to build it out. Is your grandfather still on board with our plan?"

"Yes, reluctantly. He's concerned about my safety."

"Does he know I'll be there to cover you?"

"He does."

"Well, does he have any solace in knowing that?"

"Truthfully, not much. He doesn't trust anybody in government. You know that, Pete."

The team was waiting for us in conference room 303. Eight NOAA oceanic cloud-seeding team members—a project chief, two chemical engineers, one tech engineer, one architect, and three model builders—sat around a large modular conference table. A miniature model of the simulator, ten feet by five feet, was situated on the table's center.

The project chief spoke first. "Dr. Fields, my name is Tim Callahan. I'm the project chief and I'd like to begin by saying how sad we all are for the loss of your parents. You don't know this, but those of us who were here five years ago (a few people around the table raised their hands, reverently) all adored and loved your parents. They were the best bosses we could have imagined. Not only that, but practically all of what we're going to show you today came from their imaginations and CADD models."

Pete Gerritsen interrupted. "And of course, also from Dr. Fields' grandfather, whom you folks have never met."

"Thank you all, for that," I said, holding back any emotions.

Callahan continued. "Dr. Fields, the actual simulator in India will be 328 feet by 164, slightly larger than a football field. I believe your grandfather's simulator is about the size of a basketball court unless I'm mistaken."

"That's right. And did Pete inform all of you that this is a top-secret meeting—that none of this information is to leave this room?"

They all nodded.

"Please, carry on," Gerritsen said.

Callahan continued, "In actual dimensions, the area covered in India and the Bay will be 328 by 164 miles. So, you could think of the simulator as an exact scale model of an eco-system waiting to be manipulated for the furtherance of science."

He used his pointer to show me something else about this tiny model.

"The range extends from latitude eighty-eight to ninety-two at the base and top; the longitude is nineteen to twenty-five for the width. This miniature model will help our architects and engineers explain how the real simulator will function. Our architect is up first."

"Dr. Fields, I am Lisa Chow," she nodded politely, and I nodded back. "It is my pleasure to meet you," she said, her smile looked that of an admiring student. "I read your dissertation and must say I am extremely impressed with your work." She didn't wait for a response.

She used a laser pointer. "Here, you're looking at the Bay of Bengal and here, the southeastern portion of India. As you can see, the land extends another 250 or so miles inland. This is Kolkata and our facility is up here in Darjeeling, just at the base of the Himalayas.

She pointed her beam at the blue basin. "You can see here that the Bay takes up approximately one-half of the model. The entire simulator will be enclosed with one-inch-thick acrylic walls and ceiling. Vacuum pumps will maintain a perfect positive pressure environment. We were able to fix that problem," she said, smiling. "So, what we'll have is a self-contained eco-system, to include saltwater."

I nodded. "Thank you. I'm pleased to know we won't have any pressure leaks."

Pete looked at his notes. "Next up, I believe, is Mr. Dickinson."

"Dr. Fields, I'm Ross Dickinson, chief chemical engineer. "You won't see them on this small model, but the actual simulator will have hundreds of tiny needles just above the water's surface. At the right time, the computer will release the phosphorous through these needles to super heat the surface water. These atomizers will supply a uniform flow rate of 5,000 particles per square inch at a velocity of two centimeters per second. Here's another exciting development. We've worked with GE and already at the site is a new sun lamp that will mimic the heat of the sun based on the distance from the top of the simulator to the platform. With these two heating elements you should experience real time simulations for evaporation."

"Impressive."

"Now, I will turn it over to our chief tech engineer."

"I'm Kerry Parker, Dr. Fields. I'm supplying the disc with the computations for sea water temperatures in the Bay, and the land mass temp graphs for a noon to five p.m. window when the experiments will take place. You should find these will sync chronologically with the five-hour testing time periods. That's my brief overview. Of course, later today we will fill you in on more details and measurements."

"Thank you."

"Dr. Fields, I'm reporting for the chemical engineering team. My name is Sara Jones. My job is to make sure your grandfather's discovery of coaxing ocean storms over land is successful. We'll be doing that with scale model computerized drones that fly over the land, releasing positive-charged chemicals on their way to the shoreline and the ocean." She moved her pointer back and forth over the area where water meets land.

"Coming from the ocean side are drones releasing negative ions. When the negative charged ions meet with the positive side, the cloud should move its pretty ass over land. At least it better! Please pardon my 'French'." Several of us laughed. "This is the secret part of your grandfather's creation that will draw the cloud toward land. After the cloud comes over land, we'll use his first patent discovery to seed the cloud. As the cloud moves farther inland it should be releasing droplets within 30 to 45 minutes of being seeded. Of course, we don't even know the chemical make-up he's—pardon me—*you're* going to use. I understand you'll be taking the algorithm with you?"

"Actually, no. It's too risky," I said. "Naval Air will transport it."

Jones continued, "Very well. I'll also give more detailed information after lunch."

I thanked her; then Pete Gerritsen jumped in, "Thank you all for your brief overview of the simulator. We'll break for lunch and return to this room for a walk-through of the drawings and chemical measurements."

He turned to me, "And, of course, any questions from Dr. Fields."

"Shouldn't we discuss the algorithm?" Callahan asked. I mean, since your grandfather developed it to run the experiments it's vital to our success. We've discussed this among ourselves and wish to give you a head start with the mathematical ranges—if we only knew the formula."

"Thank you," I said, "but that won't be necessary. I do not wish for any aspect of the algorithm to be part of today's overview. With the addition of my grandfather's theory of coaxing clouds over land, all the prior assumptions about an algorithm are moot. His new algorithm will include the cloud-moving part. I'm sure you understand."

Callahan nodded in agreement, but I noticed a disappointing tick from a corner of his mouth.

"If I might ask, when do you expect completion of the simulator?"

Callahan smiled and said, "Please, follow us."

We all stood and walked through a side door, which opened into a large warehouse filled with twelve large crates, each marked OF NO CONSEQUENTIAL USE—DELIVER U.S.-INDIA XCO-389162. I walked up to have a closer look at the markings.

"Dr. Fields, don't give a thought to the label. It's a diversion," Callahan said. "The actual shipping instructions are on RFID tags located under the crates."

Callahan appeared to me as some sort of government agent. *CIA? Maybe.*

"Do you mean, she's all packed and ready to travel?" I asked.

"She is," he said, proudly. "I believe it may be a record time for a project like this. But, hey, when commands come from as high up as this one did, we do what we can."

I ushered Gerritsen over to the side. "Is this new simulator they have over there a retrofit, or an original build?"

"The old one is still there. But a new cavern has been excavated to house the new simulator. All parts of the base platform have been completed—the land and the Bay of Bengal are in place. We'll be shipping these nozzles, cannisters, chemicals, and computers over tomorrow."

"When are *you* going?" I asked.

"In two days. Look, I'm afraid I need you to stay here for at least another week and see if you can help your grandfather find the dadgum algorithm."

"What if we can't?"

"Then you should come anyway. We need to begin experiments, before the Purity team muscles their way in. That's why I'm going over now. To keep an eye on them."

"Then when do you want me there?"

"We took the liberty of booking you in a week- and-a-half. Your flight info is in the procurement office. If you don't get the algorithm figured out, come over anyway. We'll find out a way to get it later. We need to start with some experiments when you get there."

"I understand. It's a good plan, Pete. Now, if you don't mind, I'd like to address the whole team and thank them."

"Of course, yes. I'll call for a meeting right away; while we're all here."

"I want to thank each of you for your hard work, the long hours you put in. You may not realize it today, but what you've done here may very well save millions and millions and millions of lives. "Thank you. Thank all of you."

Over the next week-and-a-half, I worked with Opa trying one experiment after another. Always reconfiguring the algorithm. Nothing was working.

It was time to meet Steve Muller and Dr. Bernard Loren in India.

Had I known what the next few hours would bring, I wouldn't have set foot on any plane headed to Kolkata.

– 25 –

FLIGHT 864

September 10, 2036, 6:45 p. m.

I BOARDED DELTA FLIGHT 1032 OUT OF Reagan, with a layover in London before the long leg to Kolkata. I slept for most of the overnight flight to London.

Early that morning, I boarded another jumbo out of London, Delta Flight 864 direct to Kolkata. I settled in an aisle seat ten rows from the rear. A nice-looking Black man and a teenage Indian boy came on and stored their backpacks in the overhead, one row behind me, across the aisle.

It was a full flight. Over 300.

After the lunch trays were taken away, I ran more simulations on my computer, napped some, and around four o'clock, I opened the seat back computer to look at our flight path. We were heading down and around Kolkata, out over the Bay of Bengal. I thought the flight pattern strange. Or maybe the pilots could be avoiding a storm. Not a big deal I concluded.

No more than 30 minutes later a giant lightning bolt illuminated the cabin. I looked around and noticed the Black man behind me reading a book; the Indian boy, sleeping. More flashes and turbulence woke most everyone.

Outside my window was a steady lightning show. More shaking and vibrating. This went on for several minutes—the dreaded white-knuckle ride. All passengers were awake now.

(I imagined the pilots in the cockpit bouncing like balls. "Where the hell did *this* crap come from?")

In the passenger cabin, all overhead lights powered on. The speakers crackled. "Ladies and gentlemen, I want everyone in their seats immediately. Secure your seat belts and return seat-backs and tray tables."

Only days later did I learn we were entering a 30,000-foot tornado.

I remember seeing the wing out my window being ripped off and the plane going into a roll. Oxygen masks dropped, overheads opened, papers flew everywhere. Passengers screamed, some prayed.

We were upside down, then right side up, spinning and spinning downward. I sank in my seat and gripped the armrests as hard as I could. Already in a violent dive, we were instantly headed straight down with nothing to stop us.

I remember thinking I was in a dream, in the freezing cold of the open air, seeing bodies and debris flying away from me, off into a stormy, dark distance. But I don't remember plunging into a turbulent ocean.

The Indian boy told me later that his father saw me still strapped in the airplane seat, floating, then going under. He dove in and pulled the seat with me in it to the surface. He hung on to the raft while trying to unbuckle me. He tried to lift me and the seat into the raft, but we were too heavy. He struggled to get me out of the seat but finally did and tossed me into the raft. I don't remember any of that.

The man went to work purging saltwater from me. Later, I learned he blew air into my lungs, backed off, pushed on my chest, blew again, backed off, pushed again, and gave me mouth-to-mouth resuscitation. The boy told me all this.

I vaguely remember someone saying, "C'mon lady. Breathe. Breathe."

I do remember spitting salty water onto a rubberized floor. I felt the seas rollicking, the winds howling. I thought we were going to flip over any second.

We were in a covered inflatable survival raft. The downpour from the rain pelted our roof hard. I was next to the tiny opening and could see outside. It was dark I remember; though it was still afternoon, the storm blotted out the sun.

There were five of us in the raft, best I could tell—the Black man who saved my life, two Indian boys, one elderly Indian woman and me.

I remember the Black man scouring the life raft for whatever he could find. He pulled a medical kit and some water from a bag. Then, he and one of the Indian boys began looking for other survivors among the burning oil slicks and choppy seas, yelling, "Is there anybody out there?"

Their voices were drowned by the heavy rain.

Twelve hours later I woke with the sun.

My eyes were crusted over. *Saltwater?* I rubbed and found my hand was blood red. I dared not even move for fear something else was wrong. Did I have broken bones? Other bleeding? I saw the Black man was awake and searching the sea for rescue I assumed. The injured began to wake. I remember moans and groans coming from two or was it three people. The Black man seemed to say a prayer or recite something, then gracefully slipped the elderly woman out the opening next to me.

The Black man crawled backwards to the other side of the raft, laid his head on the soft tubing, and shut his eyes. There was silence for about an hour.

Shock had set in. Acclimation was not possible. So much had happened in such a short time. I was traumatized by the plane crash, almost drowned, and found myself adrift at sea with strangers. I had seen a person buried at sea. I was facing the uncertainty of my own survival. I remember thinking that the outside world and any connection to it was gone. Forever.

My mind shifted to the groanings inside our raft. The Black man spoke first, "Here, drink some water. Eat this energy bar," he said to the boy and me—the injured boy was still unconscious. We took a sip and ate the food.

It was already hot and humid inside our little raft. I'd never experienced such body odor. Including my own.

"I blanked out," I said softly. "I don't remember anything."

"I remember everything," he said.

"Plane crash?" I asked.

He didn't respond. He kept searching the horizon.

His shirt was torn apart. I couldn't help but notice his build. Muscular, broad shoulders, perfect shape.

"Aren't you supposed to set off one of those signal things so they can find us?"

"It's called an EPIRB, and I did that last night, ma'am."

"Where are we?" I asked.

"We survived a plane crash. You need medical attention. Your forehead."

I felt my forehead and discovered nothing wet on my bandage. That meant blood had coagulated.

"India, right?" I asked.

"Not quite. We're a bloody ways from India."

"Are you the one that saved me?"

"God saved you."

I rolled my eyes. "Oh, great. We have a prophet on board."

"No. Just a missionary." He fumbled through a container inside the raft, trying to find something.

"*Just* a missionary? Well, I hope you still believe in miracles."

"Matter of fact, ma'am, I don't know *what* to believe any more."

"Don't you Christians pray to God for help?"

"Ma'am, what do you think I've been doing? Look—nobody's going to panic, okay? That's the worst thing we could do. We're in a major shipping lane. We have enough food and water, so save your breath. Stay calm and rest. That's the best thing you can do; the best thing all of us can do. Except pray."

He asked me to check on the boy under the tarp.

His right humerus was broken above the elbow—a compound fracture. Bone and blood were not what I wanted to discover. I quickly tossed the tarp back over him. I stuck my head out the opening and threw up; then rinsed my mouth with sea water.

The man reminded me, "Don't drink it. You will die."

I wiped my mouth with the back of my hand. "You'd better take a look at the boy. It doesn't look good."

I tried to relax against the sides of the raft. I wondered what sort of man he was. He was likely our best hope of survival.

He dropped the tarp back over the boy.

"What's your name?" I asked.

"Brooks. Brooks Turnage. Yours?

"Grayson Fields. I saw you and the boy get on in London."

No response from him.

Brooks must've noticed me gazing at him then back at Rinku.

"His name is Rinku."

"Hi, Rinku."

"Nice to meet you, ma'am—I guess."

Rinku looked sallow and was shaking. "Are you hurt?" I asked.

"No ma'am, just scared."

"Me, too."

Brooks broke in, "The government requires all aircraft flying over shipping lane routes to carry a three-day supply of rations for four survivors in each life raft. I checked. That's about all we have. So, I'll be rationing out food and water. Everybody okay with that?"

Rinku and I nodded our approval.

A day went by. It was hotter than ever. I was glad to be next to the opening so I could hang my head out and get fresh air.

The Indian boy with the broken arm stayed unconscious most of the time. Rinku and I were on one side of the raft, Brooks on the other side. The other boy was under a tarp in the middle. Rinku and I talked some, but Brooks kept reminding us to save our strength.

That night, I woke to see Brooks quietly and gently drag the injured boy over to my side. I scooted over to give him room; for what, I didn't know at the time. Brooks gradually pushed the young boy into the water. It was enough to wake Rinku, who saw it. Rinku lowered his head and prayed.

I quietly asked, "He was—he was dead?"

Brooks scooted backwards to the other side of the raft. Not a word about the boy he had just slipped into the ocean.

I touched my bandage to make sure it wasn't wet with blood. "Thank you," I said.

"For what?" Brooks asked.

"For the new bandage. Is it a nasty cut?"

"You'll live."

"I hope so. Did you see any other survivors? Did anyone else get into a raft?"

"Yes," he said. "I saw two other rafts with people. Maybe more. I'm not sure."

"I hope they'll be okay," I said.

"Lady, I hope *we'll* be okay," he said.

"Where did you say we are? Not that I can see anything but ocean."

"Somewhere in the Bay of Bengal. We should be in a shipping lane, so we must look for ships on the horizon. We have flares. And we have three days of rations." He rattled one of the survival cans from the bag. "The young man was dead."

I dropped my head back onto the raft's floor. "Are you going to save us?" I asked. "It looks like we're depending on you."

"We may need a miracle."

"Well then, start praying for one."

"Lady, I'm a missionary, not a miracle worker. And I'm all out of prayers."

I said, "I wouldn't even begin to know where to start praying."

Becoming tired of my sharp attitude, I suppose he decided to ignore me.

Rinku chimed in, "Ma'am, you just close your eyes and pray to God, that's how you do it."

"That's it?"

"Yes, ma'am," Rinku said, smiling.

"Even if he existed, I don't think he would listen to me," I said. "I've never prayed before. You two do the praying. I'll—I don't know what I'll do. I just can't freakin' believe what all has happened to me in the last month. I'm out of adrenaline. This is just so surreal."

"Why don't you save it for somewhere else?" Brooks said.

"What? You're one to talk. What kind of attitude is that from a missionary for god's sake?"

"Lady, I'm not an angel. I'm human, just like you."

Brooks was still angry with God. And Rinku realized that he would have to step up and be a missionary.

As promised, Brooks rationed the food and water but warned, "Look, I don't know how much longer we could be out here. So, we'd bloody well better keep calm and go slow on the rations."

"Is this all we have?" I asked, looking into the survival compartment.

"Yes," he said.

Another night and day went by, as we continued to pull the strings attached to shiny hooks, hoping to catch a fish.

Or turtle. Anything edible. Brooks would take small strips of bait and punch it through our hooks.

We took turns poking our heads through the raft's opening, pulling up and down on the makeshift fishing lines for hours, hoping for a bite. To no avail.

Then a night and another day. We had landed some small fish. Ate them sashimi, one of my favorite meals. "Got any wasabi or soy sauce in that kit?"

He didn't even acknowledge me.

"One more day of food. And we're down to one pint of water," he said.

At nightfall, Rinku and I huddled together and stuck our heads out. There was enough wind noise and small ripples bouncing off the raft to mute our voices so Rinku and I talked, while Brooks fell asleep.

I asked Rinku what it was like to live in a village as remote as Kupup. I was shocked to hear how Brooks' wife and two daughters were massacred by a warring neighbor fighting for the village's water well. I had heard of stories like this, but only on the internet.

I asked about the girls. They were like sisters to Rinku. Lynn and Lisa. Fun-loving, but studied too hard, he thought.

I asked about Brooks' wife, Anna. Rinku gave a detailed description of her, from head to toe—how beautiful she was, down to the minutest detail of her hair length, complexion, color of teeth, and especially her smile—her "warm, inviting, loving smile." Mostly though, he told me about Anna's heart. Her heart for God. "A love of God that was even more than her love for Brooks."

"Is that possible? I've never heard of such a thing," I said. "If I loved a man, he would be the only love of my life."

"Well," a calm Rinku exclaimed, "that would not be too loving to the one who created you, would it?"

"I can't see God, but I can see someone I love," I whispered, trying not to wake Brooks.

"Can you see love? No one can. But you just said you could love someone. I can't see love, but I know it's there," Rinku said.

"So, you're saying, 'Just because we can't see something doesn't mean it's not there?'"

"Yes, ma'am. God is a spirit. We are made in his image. And if you are a scientist, as you say, then you know there are many, many things you can't see but you know they are there. Like gravity," Rinku whispered and laughed at his own example.

I was surprised that I even pondered his logic for a moment. Rinku added a question: "Dr. Fields, how do you think all those stars got up there?"

We had both been staring at the billions of stars in a sky with no moon. I felt as if I were sitting higher than the horizon and the stars merely melted over the sides of my view. My heart felt something. It was deep, like a hollow chest, but with a good feeling. All of that happened in a second.

Then, I regretfully needed to think of a reply for Rinku. I was tired, but keenly interested in this young man's intellect.

"Well, Rinku, my parents taught me that an accident occurred billions of years ago, and a large explosion created life. It's known as the Big Bang."

"They didn't teach you about God?"

"My parents are—my parents were atheists. Why should they teach me about something that doesn't exist?"

"Dr. Fields, please focus on all those lights above us. Do you really believe that no energy created energy, somehow? Nothing comes from nothing. There must be something existing to make something else happen."

"I believe they are majestic, beautiful reminders that billions of years ago, a big bang occurred and created them."

Brooks' gruffy voice broke our conversation. "Get some sleep. You're going to need it. We need to fish harder and look for ships."

I glanced over at Rinku and placed a finger to my lips. "Shhh, don't wake the animals." We both snickered, sat back, and looked up at the stars again.

A minute later, Rinku whispered a question. "Dr. Fields, may I ask you something?"

"Of course, Rinku. You can ask me anything."

"If there was an explosion, and life was created from nothing, who created the explosion?"

"Well, it just happened, all by itself," I whispered.

"Well, if we're here by accident, why is there good in any of us at all?" he asked, beginning to get on my nerves.

I had to ponder his question. I should have said, "Go to sleep, Rinku," and avoided his question, but some feeling led me to believe I needed to address his question. "Well, I do think there is some good in all of us. Some just have more good than others."

"Then where does the good come from, because as you and I know, there are some very wicked people on this planet, at least according to Reverend Turnage."

"Well, I believe we develop our sense of right and wrong from the way we're raised by our parents mostly, but also from societal laws."

"But what if your parents were bad people? Would that give you— (Rinku searched for the right word . . .) *freedom* to do as you wish?"

"No, of course not. We are all held to a higher standard by society's laws and rules."

"But what if the people who make the laws are bad people?"

"We still have the laws."

"So, you believe in things you cannot see, don't you?" Rinku asked.

"No, not really. I'm a scientist. If I can't see it to explain it, I can't believe in it," Grayson said.

"We can't see the minds of those who make the laws, so why do you believe in them?" he asked.

"Rinku, you ask perceptive questions. Get some sleep."

"I can't sleep. What do you mean by per-cep-tive? What's that?" he asked.

"It means insightful."

"I think we both have faith," Rinku whispered. "You have faith that the creation of the earth and the universe was an accident, and I have faith that it was created by a supreme God who determines our life here and after we leave this earth," Rinku said, pointing up into the sky.

"What is faith to you, Rinku?"

"Faith is being sure of what we hope for and—and (he struggled to remember) and certain of what we don't see."

"That's beautiful."

"I memorized it; from the Bible, in Hebrews."

"Certainty, or as you say, 'being sure,' means there is an absolute truth in the meaning of the word. But there are no absolutes. We can only be sure of science. I'm sorry. This is probably a little over your head."

"No, not at all, Dr. Fields, I've been taught that when someone says there are no 'absolutes' they have just made an *absolute* statement."

I was quite flabbergasted by this young boy's wisdom. At least it seemed like wisdom to me. Foreign as it was. "Rinku, I've never heard it put that way. That's profound. Now go to sleep or I'll have to gag you with something."

I fell asleep watching Rinku fall asleep.

– 26 –

S.O.S.

THE SOUND OF BIRD CHIRPS WOKE Brooks right away. He knew land was close. He and Rinku stuck their heads out of the small opening and continued to search the horizon.

I squeezed my head between them. "Look, clouds," I mumbled. "Thank God."

"Well, I see Rinku made some progress with you last night," Brooks said.

"What? What do you mean?" I asked, confused.

"You thanked God. You thanked him for the clouds." Brooks was smiling.

"I didn't thank god. That was a figure of speech."

Brooks didn't argue with me, or even try to explain his point of view. He was dead tired, and not up to discussing religion. Neither was I.

We spent another day floating aimlessly in the ocean. No letup on the heat or steam.

We slept off and on, someone always searching the horizon for a ship while bobbing a shiny hook up and down.

That afternoon: "Boat!" yelled Rinku, his head stuck outside the opening.

Brooks bolted over and got his head through the opening before I could move.

"Rinku, can you tell if it's steaming toward us or in another direction?" Brooks asked.

"It's going away from us!" I shouted, peeking around Brooks and Rinku.

"No ma'am, I believe it's coming this way," Rinku said.

"Yes, Dr. Fields, I do believe in miracles, to answer your earlier question," Brooks said. "Now, help me find some flares. I think Rinku's right. It looks to be coming our way."

They located a box of flares. Brooks fired two in the air. Red and purple smoke shot up two hundred feet into a cloudless sky. We heard a pop when each flare reached its zenith. We all listened. Not sure why we were quiet. The ship was barely visible at the horizon. We waited with bated breath.

We shouted, to no avail, of course. Three heads side by side stuck out of the opening yelling for help. Was the ship coming toward us or crossing to the north?

"Don't give up hope," Turnage said, "Because if this ship is not coming our way, we're definitely near the shipping lane. It shouldn't be long now."

I knew Brooks was using psychology to keep us from being too disappointed if this ship wasn't meant for us.

"Three miles away," he said.

"How could you come up with that so fast? You can't tell the distance from here to there."

"Look, lady, one thing I know is that the ocean's horizon is a bit more than three miles from us."

Only later did I learn that Muller called his assistant. "Betty, check on Dr. Fields in D.C. She either missed her flight or is going to be a 'no-show.' I want to know which it is."

– 27 –

Senator Bennett

ON THIS TRIP, MULLER AND LOREN had been at the Purity/NOAA facility for four days. The NOAA team had finished and were there under Gerritsen's command; or so they thought.

Muller's phone rang. It was Senator Bennett.

"Yes," said Muller, answering the call.

Bennett's voice was a few octaves higher than usual: "Steve, the committee wants an update on the cloud seeding. What is Dr. Fields and Gerritsen going to tell them? You should all agree on any progress. And you've got to get the new membranes in the field, or the committee is surely going to vote for nationalization! I'm telling you, Steve, it'll get to the Senate in a matter of days. It's that hot. Get those freakin' membranes in the field! Is Bernard there? He should be on top of it."

"Loren is standing right here, Bill, and we can both assure you that our Philadelphia and Arizona plants are equipped with the new membranes. Bill, we're achieving through-put rates strong enough to satisfy congress. There's nothing wrong with the new membrane technology.

Besides, we're here taking care of a much bigger issue—the cloud-seeding problem."

"You mean you've actually gone through with that?"

"Of course. All we have to do is discredit the viability of oceanic cloud seeding and we're home free. I promise, you and your Senator friends will get that twenty-million-dollar yacht you dream about."

"But who's going to be lost in the process?"

"Bill, very little collateral damage," Muller said.

"Is Dr. Schwarzkopf one of them?"

"I don't know yet, but Grayson Fields, his granddaughter, is one of the survivors of the airliner crash. She's steaming this way now on one of our haulers. She's our way to get grandaddy over here so he can testify that this cloud-seeding nonsense is impossible. Your job, Senator, is to stall the vote. Do you understand?"

Loren leaned into the speaker phone.

"Bill, Bill, this is Bernard Loren. Tell me again how much Purity stock your family owns?"

Loren leaned back and smiled at Muller.

"Wish I'd never gotten caught up in this," Bennett said, almost under his breath.

"What did you say, Bill?"

"Forget it. But Steve, word on the street is that the membrane is not working, and this water may actually be making people sick. Something's wrong. I—I'm going to bail. I have to. I gave my confession to my priest. He says I have to remove myself from this."

"Bill! American h2O! They're the ones with the defective—"

"Sorry, Steve." *Click*. Bennett hung up.

Muller slowly switched off the intercom, still puzzled by Bennett's comment.

"What did he mean by bailing?" Loren asked.

"I don't know. But we better find out."

– 28 –

THE WATER HAULER

"SHE'S COMING STRAIGHT FOR US," I said, laughing, joyful, crying.

"Yep, sure looks like it," Brooks said.

Three hours later, "Please, may I have your names and the city where you boarded? Captain's orders," said the young Japanese ship worker to Brooks, as the three of us sat and leaned back on the ship's bulkhead, exhausted.

"I'm Brooks Turnage. This is Rinku Patel. We boarded in London."

"And you, madam?"

I was exhausted but managed to say "I'm Grayson Fields. Washington D.C."

The staffer was writing on a tablet as he walked away. Brooks put a hand on Rinku's shoulder. "I think we should get some hot tea, and fresh clothes."

"Irish coffee for me," I said. "Or a good bourbon."

"I wouldn't know about those," Brooks said.

Another staffer walked by.

Brooks asked him, “Medical help? For the lady here. Speak English?”

“Yes,” he said, looking at the cut above my eye. “Come with me.”

We descended three levels into the ship’s barebones dining hall. It was an old mess hall with worn out Formica tables, metal chairs, counters for dry goods, and a large coffee pot. The room was lit by cheap, fluorescent fixtures. It was not a modern ship.

The staffer brought in clothes, none fitting us, but no matter. I went to a corner and replaced my battered blouse and pants with a man’s shirt and trousers. Brooks and Rinku stripped down and changed. The staffer brought mugs of tea and coffee. We sat around the table, sipping on steaming mugs with the jitters of a near-death experience.

A Japanese med-tech came in and replaced my bandage. He was not at all friendly. I was still shaken and groggy but managed to raise my mug and make a toast, “Here’s to the luckiest people on earth.”

Brooks and Rinku raised their mugs but said nothing.

I continued, “I think we just survived a plane crash in the middle of the ocean.”

“Not quite middle of the ocean,” Brooks said. “Somewhere in the Bay of Bengal, near Kolkata.”

“Right. Seems like you must’ve been on this flight before. What, missionary trips, right?”

“Correct. I thought I’d told you on the raft.”

“I don’t remember. Where were you? When you were doing mission work?”

"Kupup, Northeast India, near the Bhutan border."

"I think that's not far from where I'm going—Darjeeling. So, you've been away, and you're going back there now? You and Rinku?"

"Yes."

"Well, I suppose we were never properly introduced. I'm Grayson Fields."

"I'm Brooks Turnage. I believe you know my son, Rinku."

I'm sure I looked confused because Brooks was very dark, Rinku, light brown, and obviously Asian.

Brooks noticed my confusion. "Adopted," he said.

"Right, of course. And I wonder why we never discussed this during those three miserable days in that raft."

"Because I kept us in survival mode," he said, "that's why."

"Why are you going back to India?"

"To save some villagers." Brooks stood to fetch more hot tea.

"May I ask what village?"

He laughed that laugh that spelled disbelief. "You know lady, it sure sounds like you're interrogating me. Are you with the CIA?" He turned to stare at me.

"Of course not."

"Then why are you asking so many annoying questions?"

"I don't know why. I'm just curious."

"Chhukha, Bhutan," he said to me over his shoulder.

"I don't know it."

"You don't want to know it," he said, turning his back to me.

I'm sure I looked puzzled and quite sure he could detect it.

"Have you ever seen what a bacterial paanee plague can do in a village of two hundred people?" He asked.

I was surprised by the question. "No. I haven't. I've only read about it. But what is paanee?"

"Water, in Hindi." Dubious about my presence on the plane, he said, "I'm guessing you're meeting friends for some sort of mountain climbing expedition."

I managed a smile. "You know, I'm shocked that we never discussed any of this on the raft. But no, I'm a scientist. Actually, a hydrologist."

"Really? If you don't mind my intrusiveness, can you tell me what a hydrologist would be doing in the Himalayas?"

"It's a long story."

We stared at each other for an awkward moment.

"But I suppose you'd like to hear it anyway. Right?" I asked.

"Unless you'd have to shoot us," Brooks answered, without a smile.

I smiled. "Okay, I'm a consultant for Purity. I'm on my way to the NOAA/Purity facility in the Himalayas, near Darjeeling. We believe the best chance to solve our water crisis is with oceanic cloud seeding. That's what I do."

"Cloud seeding? Why the oceans? Why not over land?"

"We know how to do *that.* Let me ask you, did you see any clouds when you were in India? Nooo, you didn't; because of the drought. The rivers, the lakes, they're all dry mud holes. And here's the whopper: the aquifers are pretty much gone. Ever heard of the Ogallala aquifer that runs underneath eight western states from South Dakota to Texas? Back in the 1980s wealthy people bought millions of acres and leases over the Ogallala. They knew water would be the most precious resource on earth. Now, they don't have drill pipes long enough to reach it. Simply put, Reverend Turnage, there's just not enough moisture over land anymore to create rain. And I'm sure you're aware that we need that rain and the snow caps to replenish our aquifers." Under my breath, I muttered, "I think you missionaries have been under a rock somewhere."

Brooks found some crackers in a basket on the counter and turned to me. "I prefer to think we've been *carrying* the rock. Anyway, look, you work for Purity. Purity Worldwide. So, I'm betting you favor privatization of water resources. Right?" He walked back toward me. "I'll bet my last Euro you know what caused the plague." He dropped the packaged crackers on the table.

"Yes, I do."

Brooks sat. "Yes, to which question?"

"To both. Yes, I'm in favor of privatization so long as it's monitored by our government, and yes to what caused the plague."

We stared at each other.

"Tainted water," I said.

Brooks raised his mug to take a sip. "So here you are, on your way to this NOAA research lab. Why the Himalayas?"

"In 2015, NOAA persuaded the administration to lobby congress and the government of India to build a cloud-seeding research facility at 8,000 feet in the eastern portion of the Himalayas, where there were clouds every day. It was the best location for experiments. It wasn't until the drought that oceanic cloud seeding entered the picture. Purity offered one billion dollars to have it retrofitted, but more secretly, they wanted to become part of the experiment team involved in all the testing. Their motive wasn't altruistic."

Brooks wanted to know, "You must believe that you can successfully create rain clouds out over the ocean and somehow draw them over land and drop their rain. Is that right?"

"Yep, that's about it."

If you're in favor of government intervention, why would Purity let you be involved? It certainly seems strange to me that they would allow you there."

"Purity wants me to fail. And if I'm there to witness it, it'll mean the end of experiments for ocean seeding. When it's all said and done, I report back to a Senate sub-committee on water resources. If I fail to make oceanic cloud seeding work, congress will vote for privatization without government oversight and let the water companies continue unabated with their filtration of black water."

"And what if you make it work?"

"Think about it. If we humans can create rain from the ocean's waters, what would the water companies have to sell? Nothing. Nil. But that's not the point. It's not about destroying them, it's about saving *us* from them," I said.

"From them? Who's them?" Brooks asked. "The water filtraters?"

"Listen, they're selling inferior body water to countries in poorer countries. Hell, even some of these bastards are selling it to American cities. Purity is the worst."

"What's his name, the guy who runs Purity?" Brooks asked.

"Muller. Steve Muller. Why?"

"Well, don't you think he'll be doing something to make sure you fail?"

"Such as, what? Give me an example. What do you think he might do to me?" Grayson knew that Muller was going to thwart her efforts, she just wanted to get Brooks' opinion on how he might do it.

"I don't know," Brooks said. "I'm just saying you should think about the consequences. You do think about that don't you?"

"No, I don't."

"But you still trust this guy?"

"I didn't say I trusted him."

"Lady, I wish I could help, I truly do. Because you're going to need it—unless—that's it, you must have the CIA backing you? That must be it."

"No CIA. No secret men in black. Just a lot of NOAA scientists."

"NOAA scientists couldn't save you if their own bloody life depended on it."

"Why are you so interested in my safety?" Grayson asked.

"I'm a missionary. It's my job to be concerned about your safety. And your soul."

"My *soul*? What's my soul have to do with anything?"

"Your soul has everything to do with it. Doesn't matter if you're successful or not. Eventually, you're going to come face to face with God. And if you don't know him, well, your little troubles here on this planet are going to seem like a head scratch compared to what happens then."

"Oh, now you're giving me the hell versus heaven pitch; but I'm not buying. Because I don't have a soul. I have a scientist's mind. And I'm tired of talking about it. Besides, my soul is none of your business."

"You know, you're right, it's none of my bloody business." Under his breath, he muttered, "It's God's business."

"What? What did you say?" I remember putting my hands on my hips.

Brooks ignored my question and walked to the teapot for a refill.

Rinku spoke up, "Ma'am, may I ask you a question?"

"Of course, Rinku, unless it's about my soul." I smiled.

"No ma'am. My question is, can you explain how this ocean cloud seeding works?"

"Of course," I said, smiling at Rinku. "Okay, well, let's see . . . oceanic cloud seeding is complicated, yet simple. Let me explain it this way. You know how clouds are formed, right? The hot sun heats up a body of water, and the evaporation from that heat makes the water vapor rise, stick together, and form a cloud. Now, this happens every day all over the world. But way back in 1949 we discovered that spraying silver iodide into a large cloud will create a larger cloud with crystals that are too heavy to remain up there in the cloud. So down comes the water." I held my hands up and dropped them to accentuate the story. "We also use other chemicals like white phosphorous. But for ocean clouds, if we don't have the exact mixture of these elements, then the cloud will dump its water over the ocean, before it makes landfall. Pretty simple, huh?"

Brooks stepped in. "But white phosphorous is dangerous. I thought it was banned long ago."

"It was. And it is dangerous. It can burn a hole in you. But my grandfather used copper sulfate and other chemicals in his patent to make the chemical combination inert but retain the heat properties."

Rinku was curious. "So how do you get the cloud to move from the ocean over the land?"

"Ahha, that's the last unknown. That's what we're working on. We've been close to solving it in our experiments at my grandfather's lab, on his simulator. But Rinku, we need this newer, more sophisticated simulator being built at the facility to test his theory."

"It's just a theory?" Brooks asked, placing his mug on the table.

"That's how science becomes reality. It starts with a theory, Reverend Turnage. If it can't be proven, then it's a worthless theory." I laughed, but they didn't.

"Is there another chemical for that? To help move the cloud?" Rinku asked.

"Yes, my grandfather has been using anions. These are atoms with an extra electron that produce a negative charge. We spray these ions over the land as we travel from inland to the ocean. Coming from the other direction, towards land, is another drone spraying positive ions. The drones will cross each other at the shoreline and hopefully draw the cloud for seeding over land. That's where we all need to cross our fingers, because we can't properly seed the cloud until it's over land. This took years to figure out. In fact, in his old simulator, we used to just spray and use fans to push the ions across the platform. Now, in India, we're going to use drones to deliver the negative and positive ions.

Once the cloud has hopefully reached land, then we spray several other chemicals into it with airplanes. The cloud usually grows dark and takes up much of the ceiling space you'll see inside the simulator. There'll be some thunderstorm-like qualities happening and then she—that's what we call the cloud—should drop millions and millions of gallons of water in 30 to 40 minutes.

"Ma'am, that's way over my head," Rinku said, scratching his shoulder. "That part about getting a cloud to move over land."

"Think of how two magnets are attracted to each other, Rinku. Opposites attract. We've seen some movement towards land, but just not all the way. It seems to stall out just before it gets to land and dump its water back in the ocean."

An awkward silence permeated the room, as neither Brooks nor Rinku had any idea what I was talking about. I wasn't worried about giving them any secrets. I knew nobody could remember all that scientific mumbo jumbo.

"I—I had no idea these things were in the works," Brooks said.

The ship's captain walked in and broke the conversation. He was small in stature, but his proud chest made him appear larger than life. His appearance spoke as that of a man in complete command.

"My apologies for interrupting. I am Captain Himashuri. You are very fortunate to be alive. The reports I received indicated your plane entered a storm producing close to 200 mile-an-hour winds. A very rare storm. I gave them your names and they said you were all in the rear of the plane. It is believed that is how you survived," finished the captain.

I thanked him and asked if I could make a call to a remote facility in the Himalayas from the ship. The Captain graciously obliged and instructed me to come up to the bridge when I felt up to it.

He turned and was about to leave, "I have one other question," he said, "if you would indulge me. We found a human arm in the raft. It had been severed at the elbow. Can one of you explain?"

"I believe I can explain," Brooks said.

Brooks told how the boy died on the second day; how he was concerned about food, not knowing how long we would be at sea. He told them how he cut off the boy's arm and hid it under the canvas and used it for bait.

"Survival instincts, I guess," Brooks said. "God provided the knife. Should I have thrown all of him overboard? What if we were still out there? Out of food."

The captain nodded one time, acknowledging Brooks' instincts.

My immediate response? "I thought that was fish bait. You know, from the fish we caught." I didn't wait for any further explanations. I didn't want to think about it anymore. I just turned to the captain and asked, "Captain, what kind of ship is this?"

A proud captain said, "We are a converted U.L.C.V. The largest of its kind."

"A water hauler," I said. "Yes, I've heard about you. Who owns it, American Water Works? Purity?"

"That is classified. But I am pleased to show you the payload."

We stood and followed Captain Himashuri to the stern deck.

It was a calm day. In the early morning light, the gigantic water bags trailing the ship were a spectacle, looking like thousands of white Moby Dicks on our tail. Looking out over the trailing water bags I could not see the last of them. Not even close to the last. Each water bag appeared to be about 30 feet wide; all were connected end-to-end, buoyed on both sides by five-foot-diameter aluminum pontoons.

We gathered to hear the captain. He explained, "She's one-and-a-half kilometers long and each bag is six meters wide. The pontoons on the sides of the bags are filled with helium to keep the bags high in the water. Designed to reduce drag."

"That's nearly, what, a mile long?" I asked.

"Almost," the captain said. "The payload carries approximately 510 million liters of water. That's about 130 million gallons for you Americans."

"I'm not American," said Brooks "Where are you taking it?"

"India."

"I hope you're taking it to Chhukha, Bhutan," said a wishful Brooks.

I turned to Brooks, "How do you know it's safe?"

"I don't," said Brooks, his eyebrows pinched together.

I turned to ask Himashuri, "Who's selling it? Avanti? Purity?"

The Captain stuck his chest out even further, "Again, Ms.—?"

"Fields. Dr. Grayson Fields," I responded.

"That is classified information, as I've said before. You should think about your good fortune. How fortunate that we were in the vicinity."

"He's right," Brooks said, as Himashuri walked away.

I ignored Brooks' and Himashuri's remarks. "It'll have a logo on it somewhere."

Brooks, Rinku and I leaned over the stern, looking at the water bags, searching for logo insignia. We soon found

Purity Worldwide's faded logo on the forward port quarter of the first bag. Small but recognizable.

"Steve Muller. I'll be damned," I said.

"Do you think any of the water companies is worse than the other?" asked Brooks.

"No, I suppose you're right."

The three of us turned back to stare at the water bags.

I turned to Brooks. "Well—it's your turn now."

"Turn for what?"

"I told you why I'm here. Why is it you and Rinku are headed to this village of yours, Chhukha, or however you pronounce it?"

There was a long, contemplative pause, as all three of us stared out over the water bags.

Brooks wasn't eager to tell the story, but he did. "I was sent to Kupup first, several years ago. We had a water well, over an aquifer. A Bhutan warlord raided our village and killed everyone to get the well."

Sadly, "I know about your wife and daughters. Rinku told me. I'm very sorry."

He continued to stare out over the mile-long trail of water bags.

"We had received word they were coming—the Bhutanese," Brooks continued, still hanging over the stern rail, staring off into the distance. "Rinku and I were on our way to stop them, to show them how to drill for their own water."

"We slept in a tree," said Rinku.

"A tree?" I asked, turning to Rinku. "You didn't tell me about sleeping in a tree."

"Tigers," he said.

"Tigers? Oh, good. A tree." I was amazed. "What happened next, when you got to the village?"

"It was my fault," said Rinku.

"*No,* Rinku!" said Brooks. "It was not your fault. It could have happened to anyone. You must forgive yourself. God has forgiven you."

"What? What happened?" I asked.

Brooks told the long, awful story, including the reason for their return to Tahir's village, Chhukha.

I listened to every word.

"Now I understand why you wanted to return. You said you saw a water truck in that village?" I asked. "I mean the wrong village you went to. Did you see any markings on it?"

"Yes. Purity Worldwide."

"Damn. Bastard."

"Still trust him?" Brooks asked.

I didn't respond. I just stared at nothing, contemplating.

Brooks broke my silence. "Tell me you're not still going there."

I perked up. "Damn straight I am. Look, he may be a greedy bastard but it's the only way I can get hold of a simulator and figure out this water problem."

"Maybe he's the problem. Have you thought about that?"

At that moment, Captain Himashuri walked in and announced, "We will reach the Port of Kolkata in nine hours. I would like to extend an invitation for you to join me for dinner in the Captain's quarters."

I spoke first, "I don't know about you, but I'm starving."

"Thank you, Captain," Brooks said, "I believe we could all use a good meal."

"Very well. I will send someone for you in two hours. If you would like to shower again or have different clothes, please just—BOOM! BOOM! Two huge explosions rocked the ship. Everybody in the galley grabbed something and steadied themselves.

Himashuri barked orders at us and pointed, "All of you! Go into the cooler room, there! In the back is a safe room. Lock yourselves in. There is a sat phone on the wall."

The ship was shuddering and trembling, and so were we.

Himashuri took off. We scrambled for the cooler. On his way out, Himashuri shouted back to us in Japanese, "kaizokutachi-me."

As we entered the food cooler room, passing hanging sides of beef, and shelves of fresh vegetables and fruits, Rinku asked, "What did he say?"

Brooks answered. "He said, 'damn pirates'."

"Wait! Wait! What if it was an engine explosion, and we're sinking?" I asked. "We could be trapped."

"Trust me. That was no engine explosion," said Brooks. "Listen—shhh. Hear it? Machine guns. The captain said 'pirates.' We're being hi-jacked."

The door to the safe room inside the cooler was hidden, and we couldn't find it. Brooks discovered a box attached to a side wall, broke it open with his fist, and pushed a red button. The door opened. We scrambled in and locked the door behind us. The safe room was only twelve feet by twelve feet. Survival equipment hung on the walls.

We could hear loud machine gun fire above us through the small air vent. Brooks found the satellite phone in a bag on the wall. He just held it, looking at it.

"Tell me you know how to use that thing," I said.

"Yes, yes, I do. But I don't know who to call."

"Give me the phone," I demanded.

Brooks handed it over. "Who are you calling?"

"Steve."

"Muller?!" Brooks asked.

"Got a better way to get us out of this? How do you work this damn thing?"

Brooks took it back, knew exactly where to power it up, opened the signal channel and asked me, "What number?"

"How did you know how to—?" I started to ask.

"Special Forces. Long time ago. The number?"

I was trying to remember the number. "Okay—let me think." *Remember.*

"Take your time. We're just going to take a little nap here, lady!"

Finally, it came to me and I spit it out.

"Could you give it to me a little slower, please," he said.

I repeated it and Brooks punched in the numbers, then put it to my ear. “It’s ringing,” he said.

“We better pray this works,” I said.

Rinku bowed his head.

“I didn't mean it literally. It’s a figure of—oh, forget it.”

“We should all pray,” Brooks said.

We could hear shouting, more shooting, then silence; followed by a barrage of bullet fire.

Finally, Muller picked up. “Are you okay?” he asked. “Slow down. Just slow down. Repeat what you said.” He listened. “Damn! Pirates. Again. Look, I'll have a plane there in less than thirty minutes. Keep calm. Look on the sat phone and give me your coordinates.”

I was in shock. “He needs our coordinates. He's sending a plane. Says it’s pirates for sure.”

“A plane?” asked Brooks. “What good is a plane going to do? We need a helicopter, or a Navy gunboat.”

I shoved the phone at Brooks. “Here, dammit, he needs our coordinates!”

Brooks grabbed the sat phone and found the coordinates indicator. “Mr. Muller, this is—never mind—the coordinates are Latitude North 21 degrees, 32 minutes; longitude, East 88 degrees, 51.4 minutes.”

“Whoever this is, hand the damn phone back to Grayson.”

Brooks gave me the phone. “He wants to talk with you.”

“Grayson, stay calm, okay?” Muller said. “I’ll have a plane out there in about twenty minutes.”

The sat phone hummed with the sound of static. We turned our attention to the vent system in the ceiling.

We could feel the ship changing course. "The pirates are in the pilothouse," Brooks said. "They're turning us around. They're Arabs. I can make out some of their words. They're making headway with the stolen water."

We slid down the wall onto the floor, sweating it out. Sure enough, in about twenty minutes we heard a jet engine coming from a distance. I stood and fist pumped. Brooks turned an ear to listen and said, "Shhh, I'm trying to figure out what sort of plane he sent."

We heard the pirates shooting at the jet as it passed over. But when we heard a much louder machine gun fire Brooks said, "Fifty caliber guns. The engine sounds like an American F-35. How the hell did Muller get one of those?"

"Government protection, and NOAA, I guess," was all I could think to say.

"Are they trying to shoot the pirates on board?" Brooks rubbed the back of his neck. He was confused. "Hell, the pirates will just scramble down here. Lord, we need your help."

The sound of the jet engine disappeared. "Gone. The plane has gone! He left us," Brooks said.

"Shhh, be quiet," I said.

We heard pirates, top deck, screaming and scrambling around. The leader barked orders in Arabic.

"You said they're Arabs. What is he saying?"

"He told his men to scuttle the ship."

"Scuttle the ship? Does that mean—"

"Yes," Brooks interrupted. "Sink the ship. Why, I don't understand. But the plane had something to do with it."

The pirates dropped explosives down hatches and vents. We heard explosions burst up through the large air vents of the ship. Then there was silence. We waited for ten or so minutes.

"Have the pirates left?" Rinku asked.

"Looks like it," Brooks said.

"But why?" I asked.

Brooks said he'd go up and find out what's going on. "Stay here," he said in a no-nonsense drill sergeant voice.

Rinku and I grabbed the survival kit and waited to hear from Brooks, but nothing.

A few minutes later he shouted down thru our small vent, "Come out. Get up here. We have a new problem."

– 29 –

The Water Bags

REACHING TOPSIDE, THE TRADE WINDS were the first sensation I felt. The other was the smell of gun powder.

We could see the pirate's vessel steaming away a couple hundred yards off our port side. Behind us, the water bags were at a dead stop, no longer attached to our burning ship, still under steam. Fire was blazing from every vent and hatch. Then we saw all the dead personnel. Captain Himashuri among them. They had been lined up and executed.

“Look! The plane shot the tow hawser off. That's why the pirates scuttled us,” Brooks said. “I'm going below to get anyone else out. You keep climbing to the pilot house and get us back to those bags before we go down!”

Brooks ran down the steel staircase, below decks, yelling to find any others. Topside, Rinku started to run down the stairs to follow Brooks.

“Rinku, don't go down there,” I yelled. “Get back up here, dammit! We've got to take this burning piece of crap back to the water bags!”

“You're going to save the water bags?!” he yelled.

I ignored his question and took off for the pilot house. I glanced back at Rinku staring down the stairwell. I called again, "Rinku, get up here now!"

I yanked the bullhorn off its cradle: "All hands on deck! All hands on deck! Brooks and Rinku, get up here now! We're about to sink. Where are you? Get up here, Rinku!"

By this time, Rinku was behind me. "I'm right here," said a timid Rinku, either frightened or astounded at my "take charge" energy.

"Yes, of course you are, you're right here next to me," I said, adrenaline beginning to subside.

As the ship closed within a few hundred yards of the bags, explosions came from midship and the stern. The engine died with a plume of smoke. Rinku climbed down to the bow, looking for Brooks. No sign of Brooks. I was still in the pilot house, on the loudspeaker, "Everybody to starboard and get ready to jump."

Brooks came running up from below and screamed, "Nobody down there." He looked up at the pilot house and motioned me to come down. It was time to jump. The ship's stern was going under and the bow was on the move upward. I quickly climbed down from the pilot house. We huddled at the bow pulpit, starboard side.

We were no longer drifting toward the water bags.

"Time to jump," said Brooks. "Rinku, you and Grayson, just swim to the bags and hold on the metal part that's down in the water. That's the hawser the plane shot off. I'll be right behind you."

We had all seen it, but Rinku was the first to say anything about it, "Look! Look at the waterspouts."

Some of the jet's bullets had hit the first bag and created two twenty-foot-high gushers.

"Everybody jump. Now!" Brooks yelled.

Together in the water, we swam to the bags and watched the ship's bow stick straight up and go under. The surface water was hissing, gurgling, smoke bubbling up mysteriously.

Getting on the water bag was no easy chore. Rinku and I eventually got to the very front of the first bag, where the hawser was hanging down in the sea and climbed up from there. Once on top it was like being on a lively trampoline, with two waterspouts raining warm water on us.

Once Brooks made it up, every move we made generated an involuntary move in the opposite direction. It became comical. Like a giant water park ball. Brooks and I sat and watched Rinku fly up in the air, plop down on his belly, fly up three feet in a different direction, and once almost flew off the bag. Brooks and I laughed hysterically. Rinku certainly had never experienced a water trampoline. He had fun like this for several minutes then settled back to reality.

"Now what?" I asked Brooks.

"Unless I'm mistaken, the current is not going to push us to land," he said.

"You have a plan?"

"I do."

"Well?"

"We need propulsion," he said. He turned to Rinku, "Rinku! Spit that water out!"

Brooks had seen Rinku, head back, catching water from the spout in a wide-open mouth. He spit it out like it was poison.

"Rinku," I chimed in, "we don't know if the water is safe to drink."

"Remember when the Captain told us the pontoons are filled with helium, to keep this thing afloat?" Brooks asked.

"Okay. So?" I asked.

He reached into the survivor kit and pulled out a knife and scissors. He gave the knife to Rinku and warned him to not drop it. They worked on the west side, the port side of the lead bag, punching holes in the aluminum pontoons. "Passshue" … "Passshue." Each time they penetrated the metal, the helium escaped with a continuous loud hiss. "Passshue." The first bag began slowly to move toward the northeast, the direction to land.

"I don't think we can do this all the way down there," said Rinku, pointing to the out-of-sight end of the bag. But they kept doing it anyway. Brooks swapped his scissors for Rinku's knife and walked, wobbly, to the lead bag's stern, telling Rinku to keep punching holes while he was going to try separating the lead bag from all the others.

We all heard the "chop, chop, chop" of helicopter blades.

"Well, at least the bastard didn't abandon us," I said.

The helo was soon overhead, lowering its wench basket. We climbed in one at a time and went up. Soon we were all aboard. As the chopper pulled away, we looked down and stared at the two giant waterspouts and could barely see the last bag almost a mile away.

"My instructions are to take you, Dr. Fields, to the facility but first drop off any other survivors in Kolkata," the pilot said, turning to look at Brooks.

"Dropped off?" asked a disappointed Brooks. "Rinku, we may have to commandeer a vehicle."

"What does that mean?" asked Rinku.

I butted in. "It means you're going to steal a car."

"Oh," replied Rinku, too tired to comment.

The thought of a missionary stealing a vehicle made me smile. I asked the pilot to call Steve Muller. I told Muller in no uncertain terms to have a vehicle ready in Kolkata.

"Why?" Muller asked. "The pilot is bringing you straight here to the facility."

"Because there's someone here, two people in fact, who are not coming to the facility. They're on another mission and I want them to have a vehicle. A new one." I hung up.

After an hour-and-a-half flight, the pilot turned to Brooks, but pointed to the ground. "Looks like your transport is ready to roll. Courtesy of Mr. Muller."

A white Land Rover. British style hunk of equipment. Brooks figured it was a 2034 Open Desert model.

He turned to me. "Thanks. Where are we anyway?"

"Twenty clicks south of Kolkata," The pilot said. "Where you headed?"

"Chhukha, up north, northeast in fact."

"You'll take N5 on the other side of Kolkata and keep north."

As the helo descended, Brooks and I looked at each other. I broke the silence.

"Well, I guess—"

"I know," Brooks said.

"Take care of yourself." I turned to Rinku, "And Rinku. Especially Rinku. You have a good son here."

Both Rinku and Brooks smiled at me. I reached across and wrapped my hand around Brook's bicep. We looked at each other as if something between us could develop.

"Special Forces, huh?" I asked. "I never did ask you back on that freighter how a soldier became a missionary. Never mind, don't tell me."

After landing, Brooks and Rinku hopped out, turned, looked back toward me, waved goodbye, and hopped in the Rover. My helo took off. As we rose in the air, I looked down wondering what would become of them.

– 30 –

HELO TO THE HIMALAYAS

THE HIMALAYA MOUNTAINS, "the abode of snow," are the tallest on planet earth. Over 100 mountains exceed 23,600 feet in height, with Everest the tallest at 29,032. The Himalaya Range extends in a slight U-shaped arc some 1,500 miles from west to east. The highest peaks are in the middle and nearer the border of China. Nearly half of the world's population lives in the basin to the south of the great mountain range.

By 2035, the abode of snow had become a misnomer. Climate change and the drought, caused by many factors, left the mountain without enough snowmelt to wet the upper regions of India and China. Entire glaciers in mountain ranges around the globe had been melting at an unpredicted rate since 2015.

Flying across India in a helicopter towards the Himalayas was a spectacular experience for me. From Kolkata to the facility was a 438-mile trip by helicopter—a two-and-a-half-hour flight for this Bell helo. The pilot was a woman, about my age. We had the usual introductory pleasantries during the first few minutes of flight. Her name was Latoya.

After we became acquainted, I had to ask, "My parents, Drs. Irene and Gray Fields were killed in a helicopter crash on the mountain. They said it was pilot error. You wouldn't happen to know anything about it would you?"

"That's BS, ma'am. I knew the pilot. He'd never had a demerit in his whole career. And, he was the picture of health, take it from me."

"So, you're saying it was no accident?"

"I'm sorry I told you that. I could lose my job."

"No, you won't. I'll make sure of it."

We settled in and flew at 500 feet and 200 miles per hour. I worked on my iPAD for an hour. That's when Latoya must've noticed that I dropped it on the seat next to me. I was now aware of the "giant ones" in front of us.

"They're impressive, aren't they?" Latoya said.

"More than impressive. They're magnificent."

"And we're still an hour away. I've been flying this route for five years now, and I never tire of them," she said. "It'll be the only thing you'll want to look at for the next hour."

And she was right.

– 31 –

Kolkata

WITHIN THEIR FIRST HOUR, Brooks and Rinku entered the outskirts of the infamous, raucous city of Kolkata. Rinku rose from a slump in his car seat for a better look, as the scene seemed to slip into slow motion.

Kolkata, the city was not what it had been in the 20th century Mother Teresa times. It was more modern—a reinvented city.

Along with the new cultural aspects of the city came the squalor. The squalor that got most of Rinku's attention was in Sonagachi, north Kolkata, home of the largest "red light district" in the world. Over 20,000 women practiced prostitution there. As he and Brooks passed through, he had his first look at unadulterated immorality.

The women began surrounding their Land Rover and brought it to a crawl. (Evidently, the women thought these must be rich patrons, since they were in a fancy vehicle.)

Rinku had never seen so many bosoms. One after another, smashed against his window, only inches from his face, block after block, for at least a mile. Dozens took their turn as the Rover rolled through the packed streets in a dusty haze. Rinku kept looking out all the windows, in shock.

Embarrassed. The women were begging for a customer. Many broke off if they spotted another vehicle.

Still staring, Rinku asked, "Reverend Turnage, how could God—" He struggled for the right words.

"How could he allow this? Is that what you're asking?"

"Yes," Rinku acknowledged.

"It's not easy to explain, Rinku, but when God created the world it wasn't intended to be like this. It was meant to be a paradise."

"Yes. And when Adam and Eve sinned, evil entered our world, is that it?" asked Rinku, staring at another bosom on his windshield.

"That's it, exactly. You said it better than I could. Ignore them. We'll get through here eventually," Brooks said.

Bangladesh was another 60 miles and had its share of prostitutes, but there were other areas in Bangladesh that made Rinku sick to his stomach. Living conditions were box after box after box of women and children living under whatever they could find to put over their heads. Outside fires with food pots emitted steam from the ramshackled huts. The smell of raw sewage and filthy human body odors was too much. Brooks flipped the air circulation vent to re-cycle, effectively shutting off the outside air. This too, went on for miles and miles.

Between cities Brooks insisted the AC be off so they could conserve fuel. The remainder of the trip from Bangladesh to Chhukha was 165 miles. They would have another three hours of uneventful driving, mixed with the sights and smells of a dozen more villages.

– 32 –

Arrival

I UNLOADED FROM THE CHOPPER NEAR a fortified concrete bunker sunk in the mountainside. I was exhausted. Latoya cut the engine and wished me well. I climbed out and noticed one other helo on the pad.

Down below, about 100 yards, I could see two objects that looked like they might be airplanes parked near a large terminal tower and a runway that must be at least two miles long. I'd never seen airplanes quite like these. Huge. Black. Sleek.

And then there were a dozen smaller planes that didn't even have windows. *They must be drones.* There was also a tram to ferry people back and forth between the bunker entrance and the airfield below.

The bunker doors in front of me were flanked by men with high-tech automatics. Two Purity guards, using earpieces with tablet screens met me.

"Credentials!" announced one guard.

A friendly Muller walked out with Pete Gerritsen.

"You should ask that man." I pointed to Muller.

Muller wanted to give me a hug. "Thank God you're safe, Grayson." But I politely held him off and turned to Gerritsen.

"Pete, it's good to see you. I hope we can talk pretty soon."

"I'll make sure of it," he said.

I turned back to Muller. "Nice to see you, too."

"I'm not sure I appreciate your tone," Muller said. "I just saved your life."

"Right. Thanks. But for how long?"

"Grayson, what happened to you back there?"

"I thought we were going to die, that's what happened. And, by the way, I've never heard of a storm like that."

We continued walking through the bunker entrance together.

"I'm just glad you made it," Gerritsen said. "You can't imagine how upset I was when I heard about the crash."

"Thanks. You didn't happen to see the millibars from that storm, did you?"

"I did," Pete replied. "It's why I can't believe you made it out alive. You were in a helluva storm. The millibar scale was 820. You were in a tornado for all practical purposes. The India Coast Guard said the plane tore apart in the air."

"I wonder why such a strong storm popped up like that. Steve, you wouldn't know of any experiments going on then, would you?"

"None."

"I assume the simulator is ready?" I asked.

"Not quite. We're working on it. There're some kinks to work out."

Bernard Loren rounded a corner and walked up.

Under my breath, I said, "Right, I'm sure there're quite a few kinks to work out."

"Miss Fields, thank God you're okay," said Bernard Loren, looking at Muller, not me. "Now, Steve, where would you like to start with our new guest?"

I turned to Steve, "Well, Steve, since we know who's in charge, I'd like a hot shower, some new clothes and some sleep. Plus, I lost my phone."

"A phone is no problem. We have plenty," said Loren.

What was I thinking? Their phone? Any phone they give me would be bugged.

"Forget the phone. I won't need it."

"Very well," Loren said, leaving Muller, Gerritsen, and me alone.

"Steve, look, you promised support for us," I said. "If there's a kink, I can fix it myself, but I need to see it. By myself."

"No problem," he said. "Come with me. I want to show you something."

"I really need a shower and some sleep first." I felt awful. I probably looked awful. But I didn't give a flying—Muller interrupted my thought.

"Grayson, this'll only take a minute I promise. Pete, why don't you come, too."

The three of us walked to a bank of elevators and descended several floors. On the way down, Pete whispered, "Grayson, you won't believe what you're about to see."

I glanced at Muller. He was smiling.

The elevator opened to a vast, cavernous, black rock cave. It had been tunneled out of the mountain. To my right, about fifty yards away was a rectangular glass building with a rounded ceiling. At the center and very top was a large mirror-like disc, that turned out to be the new hi-tech sunlamp.

The lighting from within the structure gave it the appearance of a monstrous glowing UFO. I'd never seen anything like it.

A waist-high platform consumed most of the space inside the structure. From a distance I could see what looked like mountains and miniature cities, even part of an ocean. People on the inside in white lab coats moved in all sorts of directions.

This was much, much larger than Opa's simulator.

Ten men and women from NOAA were inside working at various stations around the platform. From the outside looking in, it looked exactly like the NOAA team had told me, only this was real life, and huge.

Muller and I walked to the glass entry door, where I could see the whole operation up close. A simulated land mass covered half the platform, with water covering the other half.

"Saltwater, I presume?" I asked.

"Yes, the Bay of Bengal."

"It's all just as I had envisioned." He didn't know I'd been schooled on the entire simulator at NOAA Headquarters. Seeing the miniature trees and marshlands on the coastline made me think about those architects and engineers who built this. They were all here before me to put the finishing touches on everything, and prep for experiments.

After Muller coached us through getting our eyes scanned from a pad beside the entry door, we entered the transparent eco-system from the left corner, first through one hissing door releasing vapor, then a second door, also hissing and making sure no outside air entered the simulator.

Once inside, the view was like looking out over a realistic miniature land and ocean scape of eastern India and most of the Bay of Bengal. Underneath the forty-inch-high platform hung millions of dangling wires and tubing.

It took Muller, Pete, and me two minutes just to walk to the center of one side. Looking at it from the sidelines at mid-field gave me the feeling I was Gulliver, and this little slice of the planet was actually real. At the sideline, there was even a larger viewing stand, like a college band conductor's stand, only much larger. The three of us climbed up. Once on top, I felt like John Muir must have felt when he first encountered the Grand Canyon—goosebumps from head to toe.

I turned to Pete. "She's majestic. I only wish Grandfather were here."

Muller chimed in. "Perhaps that could be arranged."

"No, I don't think so. He's very feeble right now."

I climbed down and walked around this enormous scale model of eastern India, saying hello to each NOAA scientist I passed. I was careful to not let Muller know I had spent several hours with some of them back in D.C.

Muller followed me while Gerritsen went the other direction to engage with some of his scientists. Gerritsen had already been here several times since he arrived days before me.

"Complex, isn't it?" Muller asked.

"Yes, it certainly is," I said.

"Dr. Loren can bring you up to speed on the details," Muller said. "He's down there on the other end, working with the NOAA scientists. But I have to run catch a call. Get your iris re-scanned on the way out, for future access. Pete is already cleared. After you're finished down here, you'll find some personnel on the main floor who will show you to your room. I'll see you for dinner. Okay?"

"Okay," I said, still in a daze at the sight of the simulator. Before Muller could leave, I asked about the simulator problems.

"Don't really know all the details," he said. "Dr. Loren can explain everything. Just go see him." He smiled softly and left.

Not wanting to meet with Loren just yet, I left soon after with Pete.

On the elevator ride up, I did ask Pete about Loren. "Total jerk," he said. That's all I needed to know.

"Pete, I believe he created that storm over the Bay. You would not believe the intensity."

"Oh, yes I can. I told you earlier, you were in an immensely powerful tornado. The reports I received said the starboard wing was ripped off. I can't imagine how frightening it was for you."

"So much of it seems like a dream now. But that damn life raft. I remember it."

"Grayson, be careful. That storm was enhanced with chemicals. I saw the records. It's Loren. He doesn't want you here."

"I'm thinking the same thing. What am I gonna do?"

"Keep your head down. Watch your back. And Grayson, we'll figure it out."

"I think it just came to me."

– 33 –

Latoya and the Bird

I WAS ON MY WAY TO FIND LATOYA in the pilot's lounge; instead, we bumped into each other in the hallway. "Shh, come with me," I said, taking her by the arm. "I was coming to get you."

She started to resist but stopped. "What's going on, Dr. Fields?"

"Shhh. I'll tell you in a minute. Just come with me." We entered a restroom. I searched to make sure it was clear.

Latoya was confused. "What's going on, Sugar. You can tell me."

"Latoya, you know cameras and bugs are everywhere. We have to be discreet."

"Okay, what's up?"

"I have an important question for you. How well do you know Dr. Loren? I ask because I'll be meeting with him sometime tomorrow."

"Well, let's see, I've flown him over here and back to the states—let me think—yep, three times now. He likes to sit right behind me to my right. He wants to see everything ahead of us. Naturally, I hear everything he says on his phone. He doesn't realize I can hear him plain as day through my headphones. He's a prick as far as I'm concerned. I've heard him give orders to eliminate people."

"You mean fire them?"

"Hell no, eliminate them! You know . . . finito. I dated an Italian guy once and he—"

"Let's go have a cup of coffee, shall we?"

In the cafeteria, I told her, "We have to be careful. Everybody in the facility is being watched."

"Sweety, I spend most of my time in the pilots' lounge. We have everything we need. I just stay outta their way up here."

"Latoya, I'll be straight with you. I need your help. I need to get back to the states. I believe my grandfather may be in grave danger or will be soon."

"Your grandfather has something to do with this facility?"

"Yes, he does. He holds the key to these experiments. He's developing the algorithm that'll make it work."

Latoya just pinched her lower lip and stared at me for a few seconds.

"By the way," I asked, "how long is the flight from here to Washington?"

"A little over four hours."

"That's all? Four hours?"

"Did you see the interceptor planes down the hillside?" Latoya asked.

"I saw two black objects that looked like planes."

"They're the new FX-40 HyperSoar RamJets. They cruise at 6,200 miles per hour."

"You've got to be kidding. A regular airliner takes eighteen hours. How is it possible for you to—"

"Do it in four hours?" she completed my thought. "Because these are military. These planes have no respect for time and distance. There're only twelve birds like this in existence. NOAA has had two of them here for ten months. It's beautiful up there, above the earth."

"So, you'll do it?"

"Grayson, I like you; but I'm gonna need to give this some thought."

– 34 –

Chhukha

TAHIR WAS ON A MAT, PROPPED up against a dead palm tree, near the village hospital. He heard the hum of a distant engine. He turned to see a vehicle coming his way, dust swirling all around it.

As Brooks and Rinku arrived, they slowed to a crawl and could only gape at all the bodies. Men, women, children, all dead.

They crept through the village looking for any survivors. None. The flies were merciless.

"Ohhh, my God. My God," Brooks said. "I had no idea so many would be—gone. They're all dead. Oh, no, so many of them, hacked to death like this. It's not right."

Rinku leaned forward almost touching the windshield. He could see some villagers on their backs, spears embedded in their chests, handles pointing to the sky. This brought back horrible images from the hills above his old village.

Some appeared to have either died from a plague or had it when the village was overrun by some warring tribe.

They kept moving, slowly, quietly counting the number of dead as they passed by. ". . . forty-six, forty-seven, forty-eight."

"Rinku, I'm afraid he's gone. Maybe they took him."

"No, he's alive. He's gotta be. I've heard old stories about how Circle Leaders have an agreement—"

They were half-way into the village when Rinku spotted Tahir. "Look! There." he shouted. "Leaning against the tree."

"It's Tahir. I see him," Brooks said.

Brooks rolled to a slow, reverent stop. Both calmly opened the doors and stepped out. In shock.

Brooks shouted, "Get back in!"

Rinku scrambled back in the Rover, startled, confused.

"Rinku, cover your mouth and nose. I don't think the plague is airborne but do it anyway."

Rinku tore off the tail of his shirt and wrapped it around his nose like a robber's bandana. Brooks pulled his t-shirt over his nose, then turned to inspect Rinku's protection. Satisfied, he poured water from their bottle on Rinku's new mask, and then some on his makeshift mask.

He stepped out of the Rover. Rinku followed, watching every move Brooks made. They slowly walked over to Tahir. Brooks knelt beside him. There was an old canteen beside Tahir. Brooks picked it up. It was empty.

"Is anyone—ask him if anyone else is alive," Brooks said. Then Brooks changed his mind, pointing back toward the Rover. "Rinku, no. First, get some water from the truck."

Rinku returned with a bottled water.

Brooks took it. "Go, search the village. See if you can find any other survivors."

Rinku ran. Brooks yelled, "Keep your face covered."

Brooks poured a small stream of water into Tahir's mouth. Only Tahir's eyes moved—to look at Brooks. Tahir swallowed and prayed something. But Brooks understood none of it.

A few minutes later Rinku returned. "None. Not one." Rinku was crying.

"Rinku, did you search the hospital?"

"There was a large tent. I looked inside. I couldn't stand the smell. I'm sorry."

"Okay, stay here. Give him little sips of water. I'm going to take a look. There's somebody I want to find. If possible."

The tent was 100 yards away. Brooks pulled his t-shirt tighter to his nose and walked through the same tent Madan had shown him. Most of the dead seemed to be victims of the disease. He had to step over several bodies to reach the exit on the other side. He found Madan outside on the ground. She hadn't died of disease. She had a long wooden spear in her chest. Her eyes were still open. Brooks knelt, closed her eyes, and said a prayer.

He returned to Rinku.

"Rinku, I expected there to be more alive. They're all gone. I guess we just need to take Tahir. I wonder why they left him alive. Why didn't they kill him, too?"

"I started to tell you just before I spotted him. I've heard Manu's family talk of it. Several times. The Bhutanese leave village chiefs alive because they believe bad spirits will come on them for killing a chief."

"You didn't tell me, Rinku. Did they leave Manu alive?"

"I never saw." Rinku's shoulders slumped and curled forward. "Remember, I crawled out, like a coward before I saw everything." Rinku's eyes had misted over.

"Rinku, I've told you a hundred times, there was nothing else you could have done. Look at me, Rinku! I would have done the same thing. Now we need to get the hell outta here. Go let the back seat down. We'll lay Tahir in the back."

"Where are we taking him?" Rinku asked.

"Somewhere near the Himalayas. The NOAA facility. Best chance he'll have to survive." Brooks moved his eyes close to the nav map on the dash. He was not doing anything, just staring at it.

"What's wrong?" asked Rinku.

"Rinku, when we were on that freighter, Dr. Fields mentioned where she was going. It's a city or a place where the facility is located. Can you remember it?"

"No sir."

"Think, Rinku, think." Brooks took his foot off the brake and they rolled forward several feet. Brooks placed his head on the steering wheel trying to remember the name."

"I believe it was something like darling," he said. "I remember thinking of Anna when I heard her say the word."

"There is a city named Darjeeling," Rinku offered.

"That's it!"

Twice, Brooks entered the wrong spelling. Finally, he entered Darj and the map went to the route. "There it is. Darjeeling. God bless you, Rinku. It's only 143 miles away."

Brooks hit the accelerator and peeled out, barreling the Rover over the one-way dusty roads at eighty miles per hour, listening to Tahir's occasional prayer, but understanding only the parts Rinku tried to translate.

Brooks began thinking of how he was going to be able to enter a secure facility if it was indeed anything like the one she talked about. *It's going to be heavily guarded. In the mountains, one-way in. How am I gonna get us in?*

– 35 –

Brooks at the Gate

AFTER THREE HOURS, BROOKS AND RINKU—with Tahir, half dead—pulled to a stop at the Purity/NOAA check point, where Brooks encountered a bright halogen flashlight in his eyes.

"Credentials," the armed guard demanded.

"Look, soldier, I don't have any credentials on me. They're in the Bay of Bengal right now. So, I need for you to call Mr.—"

"Turn your vehicle around and leave immediately," the guard demanded, not even interested in hearing who the "Mister" might be.

"Wait. Call Steve Muller. I have his cell number if you don't believe me. Here, just call him on this satellite phone. I've put the number in. He'll approve us."

The guard stared at Brooks for a couple of seconds, checked out the unhealthy passenger in the rear again, then crisply walked to the guard booth, leaving the sat phone with Brooks. Brooks sensed he'd be denied entry, so he stomped the gas pedal, trying to break through the gate. The gate pole bent backwards a few yards, then popped up over the hood and remained stationary on the windshield. Stuck.

The soldiers manning the guardhouse quickly lowered their rifles on the vehicle. The guard came out in a fury, pointing his pistol squarely at Brooks' head. Brooks' satellite phone was already at his ear.

The guard shouted: "Put your hands over your head, mister, this is a secure United States facility."

Brooks and Rinku both raised their hands. The phone was above his head. "It's ringing now," Brooks said.

"Give me the phone." Brooks handed it over.

The guard put the phone to his ear. He heard Muller's voice: "Who is this?"

"Who am I speaking to?" asked the guard.

"This is Steve Muller. Who the hell is *this*?"

"Sir, I'm the guard, at check point one. There's a man and some others down here who claim to know you."

"What's his name?"

The guard looked in the open window at Brooks, hands still above his head, "What's your name?" he asked.

"Brooks Turnage. Reverend Brooks Turnage. I'm a missionary out of London."

Muller said, "I heard. Hold for further instructions."

Muller turned to Grayson, "There's a man named Turnage at the front gate. Wasn't there a Turnage with you on the plane? And the ship?"

"Yes, why?"

"He's outside the front gate, right now."

"Let him in!"

"Bull. Why should I let him in. What's going on, Grayson? Why's he here?"

"He's the missionary from the plane crash for god's sake. Maybe he needs medical attention, I don't know."

Muller eased the phone back to his mouth, "Guard, do any of them look sick or injured?"

The guard leaned into the back window and asked, "Any of you sick? Injured?"

Brooks' eyes widened in disbelief, "What the hell do you think he looks like? He's an old man and he's sleeping. But he does need medical attention, so let us in!" Brooks shouted into the guard's phone.

The guard answered Muller's question, "Sir, there's an old man in the back. He's sleeping but I'm not—"

Muller didn't wait for him to finish, "Check the vehicle for weapons. If you find so much as a pistol, I want you to send them on their way. Do you understand?"

"Yes sir."

Grayson heard every word. And moved in. "Weapons? Steve, he's a missionary for god's sake. And he doesn't carry a pistol. Now let them in!"

"And what? You think just because he's a missionary he doesn't work for the CIA, or MI6, maybe coming to infiltrate our operation? They did it in Lagos, but you wouldn't remember that."

"Look, Steve, I know him. He's a missionary."

"He's no missionary. He's the one who called from the freighter. He's military. I'm sure of it. Did he not tell you that?"

"Of course, I knew! But he's been a missiona—"

"I don't care, Grayson. We're holding them until I hear from the guard."

The guard came back on the phone, "No weapons, sir. What should I do?"

"Tell 'em to—go back where they came from. Tell them we're not a hospital. Something's not right about this," Muller said. He tossed the phone on a desk.

I shook my head and walked back to the semi-circle row of super-computers. Muller followed me. I was tired and frustrated. "He doesn't need a gun," I told Muller. "He could kick your ass with his bare hands."

"Whoa! Wait a minute, young lady. What's come over you? I don't know what it is, but it's not good."

I sat down at the terminal. Exhausted. I lowered my head and screamed, "Steve, just let the man in."

"Grayson, what is it? Tell me."

"He saved my life. You could at least afford him that."

Muller spun around, aggravated, thinking, wondering what to do. Finally, "Okay, I'll trust you, Grayson. I'll let them in. But mark my words, if he causes any trouble, I will have him eliminated."

Eliminated? That was the first time I'd heard him mention anything about eliminating someone. I had to keep my cool. Keep playing my part. I had to remind myself that acting had gotten me this far. I couldn't quit now. I needed to buck up and face the one who would rather starve half the world if necessary, just to maintain his reputation as "World's Richest Man."

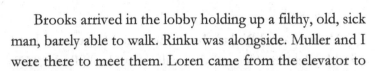

Brooks arrived in the lobby holding up a filthy, old, sick man, barely able to walk. Rinku was alongside. Muller and I were there to meet them. Loren came from the elevator to join.

"So, you're Brooks, the missionary who saved Grayson's life?" Muller asked.

Brooks nodded. "We could use some medical attention," he said, glancing at Tahir.

"Berney, have someone take them to the infirmary, would you?"

Berney's scowl towards Muller plainly showed he didn't care to deal with someone so filthy. Still, he escorted Brooks and Tahir down a hallway to the infirmary, leaving Muller and me with Rinku.

"Sir?" Rinku asked, "you're Mr. Muller, is that right?

Muller merely tilted his head down and said, "Yes. And you are?"

"My name is Rinku. Reverend Turnage is my father."

Muller put his hand on Rinku's shoulder. "Well, Rinku, glad to meet you. So Reverend Turnage is your father?"

"Yes, sir. Best man in the world."

"I'm sure he is," Muller said. "I understand all of you had quite an ordeal back there."

Rinku looked over at me. "Yes, sir, we did. Dr. Fields and I prayed a lot."

Muller's shoulders flew open and his eyes grew wider, "Dr. Fields and you prayed a lot? Is that what you said?"

"Yes, sir."

Muller patted Rinku's back, then tried to hug me, but I wouldn't let him. He turned to walk off and told Rinku, "Dr. Loren will be back in a minute to take care of you. You're safe now."

Muller gave me a little evil eye, then strutted down the hallway to meet Loren, on his way back from the infirmary. I couldn't hear what they said to each other. But Muller turned around and came back.

I gave Rinku a hug. "Are you all right? What happened? Who is that man you and Brooks brought with you?"

Rinku's big grin almost made me cry. "Thank you, Dr. Fields, I am fine. I should let my father tell you what happened. I'd like to see him now."

"Come with me. I'll take you there."

Muller returned. "Young man, you look like you could use a bath. Come with me. I'll show you where you can shower and change."

"Thank you, sir, but I'd rather see my father if you don't mind."

I put an arm around Rinku and said, "I'll take him down. He wants to see his father first."

"Good," Muller said. "I actually have an important call to make. I'll catch up with you later."

I knew he and Loren would be checking out Brooks. They'd be calling someone in the states. Find out all about him. That's what they would be doing. I just wish I knew who their contact was.

At the time I had no idea that Loren would also be giving someone instructions to eliminate me.

Rinku and I walked into the infirmary. Brooks was sitting on the edge of a chair, leaning forward, the palms of his hands covering his face. He was beside Tahir's bed; both were praying, Brooks in English, Tahir in his native tongue. Brooks didn't notice us walking in. We stopped and stood quietly by the door.

". . . and so, Father, I know your promises, and I know you always and forever keep your promises. So, I ask again, no I beg, Father, if there are ten people here who are your children, will you allow them to perish along with the wicked? No, you will not! You love your children, and you protect them."

At the same time, Tahir was praying. I noticed something I hadn't experienced in a long time—chill bumps—up and down both arms.

"We, your children, are dying of starvation," I heard Brooks say, as I hugged Rinku tighter. Brooks was still unaware that we were in the doorway. "There is not enough water," he prayed. "This place we're in—this facility, is a place for helping make rainwater. Your people need this, Father. You know me. And you know Rinku. Dr. Fields doesn't know you, but her work is what I ask you to bless. Show her the answer to rainwater. And, Father, show her the way to you. Create in her a new heart; a heart that desires to know you, and a soul that loves Jesus."

As he kept praying, I noticed another sensation. An unexpected tear slipped from my eye. My heart was filled with the same majestic feeling I'd felt in the helicopter as we flew closer and closer to the Himalayas. A feeling of wonderment.

"In His precious name, I ask these things. Amen," Brooks said.

Brooks rose from his knees beside Tahir's bed (who never quit praying) and turned toward us. "I didn't see you come in."

"I'm sorry we interrupted," I said.

"No, not at all. I was just—"

"I know. Praying. I heard. And Brooks, thank you."

"For what?"

"For praying. Praying for me."

– 36 –

NEW PAJAMAS

AFTER RETURNING TO MY QUARTERS ONE night, I discovered a pair of pajamas neatly placed on the bed. I figured it was a gift from Muller. One of his charming acts to coax me into *his* bed.

Too tired to think about it, I enjoyed a long hot shower. Afterwards, I slipped on the pajamas and stretched out on the bed. They felt nice. Silky. Almost slick.

About fifteen minutes later I felt some strange sensations in my chest and back. I began scratching. Vigorously. It made me feel worse. Nauseous.

The material was beginning to stick to my skin. I remember feeling woozy. Sweat seemed to be coming from every pore in my chest and back. Instinctively, I tried to raise my top to smell whatever it was making me sick. But my shirt was stuck to me like glue. It wouldn't budge. Something was terribly wrong. *Think*, my brain screamed. I had never felt sensations like these.

Tripping over a chair, I fell but managed to crawl to the bathroom shower. I was gasping for air. I could hardly breathe. I turned the shower on and sat. After a minute of shower water pouring over me, my shirt loosened up a bit. I was able to rip it over my head and peel it off.

I watched a bright yellow chemical go down the drain. I stripped down and soaped and scrubbed every inch of my body for five minutes. I dried off, then took a pair of pants from the closet and ran sink water over the hem of one leg for several minutes watching for a yellow chemical. Satisfied that the pants weren't a problem, I repeated the process with one of the shirts.

I dressed and wrapped a towel around my hand to act as a glove, picked up the tainted pajamas, rolled them up in another towel, and took it all to the garbage slot in the hallway where I shoved them down the chute. I rushed back to my bathroom and force vomited. Still woozy, I stripped off all my clothes again and stepped back in the shower. The yellow chemicals in the shower were pale now and gradually morphed to clear water.

I dressed again and found my way to the cafeteria, grabbed a pint of milk, drank it all in one long gulp, and headed back to my room.

Still sweating, I was acutely aware that I could have died. I decided to keep this to myself, at least for now. This was the second, maybe the third time someone tried to kill me. *It's gotta be Loren.*

Exhausted, I tossed and turned the whole night. I kept asking myself what I could do to get out in front of what was surely a bad place to be. I needed to get ahead of whoever it was trying to kill me.

I could go to Pete and ask for his help. I could go to Brooks. But what could either of them do to keep me alive?

What I needed was for the experiment to happen as soon as possible and to happen successfully. What I needed was the algorithm. I needed Opa.

– 37 –

TURNAGE'S FILE

THE NEXT MORNING, LOREN WAS back in the security control room with Muller.

"I did a little checking on our missionary guest, Turnage," Loren said. "Seems he was born in London, joined a punk gang at sixteen, sentenced to juvenile detention at eighteen, then joined British Special Forces. Awarded several medals of valor, retired at 34, and became a missionary in India."

"Is he a spy?"

"I don't believe so. He's been in some remote India village for five years. Has a wife and two daughters. Best part is, he's no longer classified."

"Thank god. We don't need another problem. Where's Grayson?" asked Muller."

"Sick bed, I heard. She'll be okay. Doc is checking her out now. Nothing serious."

One of the monitors in the security room was devoted to closed circuit TV news. When Muller saw the caption on the screen, WATER AUTHORITY VOTE NEAR, he turned the volume up. They both listened.

The anchor, George Gwin, was delivering the day's news.

"...and *that* from our correspondent in Spain—again, one of the worst-affected regions of the world. We take you now to our correspondent in Texas, Bill Foose. Bill?"

"Yes, George, I'm here in Lubbock, Texas, and as you can see around me, it's utter chaos. The drought here is being compared to the great dust bowl of the 1920s. There simply is no water. The exodus of people, families, and businesses has brought this Texas city, and many, many more like it to its knees. So many people, George, are telling me they're moving to the Rockies, and even Canada, in search of water, and hopefully, cooler air. But the hope of finding plentiful water even in the north is still a gamble for these Texans. It's a sad situation here in Texas, George."

Gwin: "Here now is E.J. Baird reporting from the Capitol. E.J., what have you heard from the Water Resource Committee? Will they recommend a bill for nationalization?"

This is the part that Muller and Loren had patiently waited to hear.

"George, I'm here on the steps of the Capitol," Baird said, "and yes, the latest word is that the Committee will approve the bill to go forward with a vote for nationalization. There's another bill to follow about price controls which will come up for vote after this bill is decided. More details on that possible bill are to come. But for now, it appears the big bill will go to vote in the Senate next week; and as you know, the House has already passed *their* version, so we could be in for more government control of our water. Back to you, George."

Muller slammed his hand on the console. Loren lowered and shook his head.

They looked up as Gwin came back on. "This breaking story just in from our Washington bureau. Senator Ray Gill, the Arizona Republican, was found in his Washington apartment this morning, apparently dead from a heart attack. More details at ten tonight."

Muller, showing no emotion whatsoever to Gill's death turned to Loren, "I can't believe they're going to vote for nationalization. We could be smothered to death by a bunch of bureaucrats sticking their noses straight up our butts and dictating our prices. We'll be screwed."

Loren knew what needed to be done. "We have to get the simulator up, and we need to make at least six runs to show the idiots they can't count on ocean cloud seeding. Congress needs to know they need us."

Muller had something else on his mind. "You don't think Gill got to the media with those photographs, do you?"

"My people knew to burn the photos. I got confirmation of it. Even if he made copies for the media, we'll rebut everything, every step, every inch of the way with your story: American h2O stole several of our tankers and put their tainted water in all of 'em. The police report—it's already on the books; and our judge, you remember Henry Reddick, don't you?"

"Yeah, an old pal of my father."

"Right, he has the case. But he liked your daddy so much he's gonna do us a favor and go with heart attack on the death certificate. The public will buy it."

Muller grabbed Loren's shoulder, squeezed and said, "Good."

– 38 –

The Bar with Latoya

I WAS TIRED AND DECIDED TO FIND the lounge. The facility housed a large cafeteria with an adjacent classy oak-paneled lounge. I sat alone at the bar and ordered a French Martini, something I hadn't had in a long time.

"Sorry, mum, haven't had a request for that one in a while. Afraid I don't know the recipe."

"A half jigger of Chambord, one jigger of pineapple juice, a jigger-and-a half of Chopin vodka, if you have it," I said. "Otherwise, Tito's from Texas will do fine. You know to shake it vigorously, right?"

"Yes ma'am, that's when a soft pink foam will form at the top, right?'

"Exactly," I said.

Well, I had two of these before my new pilot friend wandered in, looking for company I suppose.

"How you doin', Sweety? You get settled in okay?" She asked.

"I—I did, thank you."

Latoya turned and gestured to the bartender, "Bourbon on the rocks." She looked at me. "Don't worry about me Hun, I'm off duty for forty-eight hours."

We talked for a few minutes about the Himalayas and their mysterious beauty. She showed me photos she'd taken from her helo a few years ago.

"Look at these. See? It was so majestic back then. I mean they still are, but just a few years ago they were covered in these snowcaps. You should have seen them then. My goodness, they were really breath-taking."

"When you brought me up here, just looking at their majestic beauty gave me goose bumps. Mount Everest, right there in front of me."

"Speakin' of goosebumps, where's that handsome Brit who showed up here with the boy?"

"Ha." I laughed. "I'm pretty sure he's not a drinker. He and the boy went to a remote village in Bhutan to help but they found only one person alive, the tribal chief. He brought the man here for medical attention."

"Strange. How did you know him?"

"He and I, and the boy Rinku, happened to be on the same flight, the one that went down in the Bay of Bengal several days ago. He saved my life."

"Whoa. You're kidding me. Right? I heard about that. Just a dozen or so survived."

"Wait, did you say a dozen or so survived?" I wasn't aware how many had made it on the other rafts.

"Yeah, that's what the news said. The Indian Coast Guard picked up several rafts. And oh, my Lord, I'm sitting next to a survivor from a crash that killed three hundred people? And he saved your life? You've got to be kidding. Bartender bring us another round. Now tell me, what happened? The crash, I mean."

"Well, let's see. I guess I remember seeing bodies flying away from me through lightening, rain, just going off into the distance. That's the spookiest thing to me about the crash. I heard our plane had torn apart and the tail section went into the ocean backwards. That's what I heard. I was in the ocean, sinking, still strapped in my seat. He dove in and pulled me into a life raft."

"Who did? The Brit?"

"Yes. 'Course, I don't remember much about the crash. We spent three days and four nights in that raft." *Hic.* I suddenly had the hiccups.

"What's the man do?"

"The man? *Hic.* Oh, he's a missionary."

"He ain't no missionary, darlin', he looks like a bodybuilder to me."

"He used to be British military—*hic*—something-or-other."

"I knew it. I knew it!" She said. "I was trained by the Navy. I can spot one when I see one."

Our drinks came. It didn't take long to go down. *Hic.*

"Honey, here, take this glass of water and drink it upside down."

"Who filtered this water?" I muttered.

The bartender said, "Who do you think? Purity."

I pushed it back and waited for the bartender to walk away. "Latoya, didn't you tell me earlier that your helicopter has a way to communicate with America from here?"

"Yes, for sure. Why?"

"Is it secure from all the intercepts—*hic*—in this facility?"

"Yes, it's secure. It's all in the helo. What's up, Dr. Grayson? You need to make a private call?"

"My grandfather, that's all. I need to check on him."

"Well, because I like you, I'll set you up. But first we're gonna get some coffee in you. Bartender, pour Dr. Grayson here a cup of hot coffee?"

The bartender went to work on the coffee machine. Latoya said, "Look, Hon, if I were you, I'd wait for those three martinis to wear off before I called anybody, especially my grandaddy."

"*Hic*."

After one cup of coffee, she took me out to the helo. "Who knows, we may not find another chance to do this," she said and made the call. She explained how to hang up after I finished. When Opa came on the line, Latoya gave me a thumbs up and climbed out.

"Opa it's me, Grayson."

"Grayson, I knew it would be you! Are you okay?"

"I'm fine. Tell me where you are."

"I'm at the lab."

"Good, have you made any headway with the algorithm?"

"Oh, my little Grayson, you wouldn't believe it. We made a giant one with a new formula of mine—"

"What do you mean 'we.' Who was there?" I was afraid it might be Purity spies.

"My grad students! Anyway, she was huge. We sprayed the positive magnetic ions over the water just as I taught you, then used the ground sprays for the negative ions over the land. Grayson! She moved! She came to land."

He was so excited I was afraid he might have a heart attack.

"Oh, but Grayson, just before she reached Manhattan, she unloaded more fresh water than I'd seen. Naturally, we tested the specimens, and they were as pure as rain. Oh, if only she could have made it to land." I could hear the thick German accent of disappointment in his voice.

"You must be close, right?"

"Yes, tomorrow, I know what to do. I will tweak the algorithm and seek to move her all the way to Albany! We should have the algorithm soon, my dear child." He sounded so confident. Again, I shed a tear. So not like me! *Gotta be the martinis.*

"Now," he said, having recovered from his exuberance, "tell me about that simulator where you are." He seemed excited to hear about it.

"Opa, you won't believe it. I wish you could see it. It's your dream come true. It's five stories underground. It is indeed larger than an entire football field. It's perfectly sealed and has its own replica of our eco-system. The ocean portion is even curved, like our earth. They say it's some sort of anti-gravity mechanism, something you and I have never seen before."

I went on to describe the whole operation.

"How many experiments have you done?" he asked, his voice two octaves lower, "although I know you couldn't have been successful without the right algo," he added.

"I witnessed one," I said. "And they ran two before I arrived. And you're right, Opa, none worked. But Opa, once you have the algorithm, you'll give it to Jesse, right?"

"Dear, don't be alarmed, But I'm going to bring it over there."

"Oh, Opa, please don't. Is something wrong with Jesse?"

"No, nothing's wrong with Jesse."

"Grandfather, Opa, please listen to me. It's just too dangerous here. Trust me. I know how to get back there within four hours. I'll bring Jesse back here with me and go from there. Okay? Please, Opa, promise me you won't try to come here."

"My dear darling, Grayson. I will do as you ask."

I didn't believe a word of it. He's too bullheaded to give in that easily. We ended our talk with soft goodbyes.

As I reentered the facility at 2 a.m., Steve Muller was waiting at the entrance. "Grayson, what are you doing out here? There's nothing out here except the guards and the helos."

"Taking a pee. I like it outdoors."

He smiled. "But the guards, they were watching you. Weren't you embarrassed? Have you been drinking?"

"They didn't seem to mind. Besides, I was out of their site. You know, Steve, I've been thinking."

"Me, too, Grayson. And I think it's time—"

"Wait, me first. I've been thinking that you had something to do with my parents' helicopter crash five years ago, didn't you?"

It took me a while to get all that out, but I managed it.

"Grayson, Grayson, don't be so naïve. You know I could never do something like that."

"Oh, yeah? How could you *not* know about the storm? And that I was poisoned? Huh?"

"I only learned about that the other day. I'm just thankful you're okay. Grayson, listen, you and I . . . we're on the same team. Speaking of which, you remember what you promised me back in D.C. don't you?" he asked, smiling.

"What?"

"You said you'd sleep with me in this exotic land. You remem—"

"Don't be a moron, Steve. You're not that attractive to women. At least not to intelligent women with taste."

Right then and there I regretted having had too much to drink. I should have continued to play along. Now I had really pissed him off.

– 39 –

Black Ops

MULLER CONTACTED BENNETT IN D.C. via back channels.

"Bill, you must get the algorithm from Schwarzkopf. Do whatever you need to do but get that algorithm back here to me at the facility. We need to erase it. Destroy it. After you get it, demo that whole damn simulator of his. And him with it."

Bennett didn't want to do it but concluded that he was in too deep. The next day he made a call to his ops contact. After giving him the instructions, he called Muller.

"It'll be done in a week."

Bennett leaned back in his Senatorial office executive chair and sighed deeply. "What have I done?"

– 40 –

Hoya Gym

ON THE ROOF OF THE OLD HOYA gym, a man dressed in black drilled a hole into the gym's ceiling next to a large roof vent. He pushed a tiny camera on a cord through the hole and sealed the area with caulk. He then taped a backpack inside the vent. He set a clock inside the backpack, placed a tiny antenna on it, zipped the bag up, and repelled from the building into a patch of dry grass.

– 41 –

Another Call to Opa

TWO NIGHTS AFTER OFFENDING MULLER, I snuck out again and called Grandfather. He didn't answer.

I called again the next night. No answer.

Next morning, I visited Brooks in the infirmary, sitting with Tahir, who never stopped praying. "I forgot to ask you what his name is," I said.

"His name is Tahir. It means holy one. He's the chief, or was the chief, of a village named Chhukha. I'm not sure which tribe decimated his village for the water. But it doesn't matter. It's not customary for the people in this part of the world to kill a village chief. They leave him to watch his people die. He was half dead when we found him. He hasn't stopped praying since we left his village."

Brooks turned back to look at Tahir.

I thought I'd never heard that many words from Brooks Turnage since I met him.

"Who does he pray to?"

"He prays to the one and only God of our universe, God the Father, God the Son, and God the Holy Ghost."

"You're speaking of the trinity, aren't you? I'm beginning to believe you, about this God of yours. In fact, we may need another miracle."

"How so?"

"Look, Brooks, I'm concerned about my grandfather. I told you about him. He's never been without his phone. He hasn't answered in two days. Please, come help me find him."

"How?"

"Sneak away and go with me to Washington. I need to find him."

"What? That's impossible. He's on the other side of the world."

Tahir stopped praying and slowly turned his head toward us to reveal a face of a thousand wrinkles and dark, deep-set eyes, with a graceful, easy smile, signifying that he would like some quiet so he could continue praying. I thought to myself, *goodness, the man could have shot us both an ugly stare, but instead, he gave us a lovely calm smile; and we both got it.* We moved to the hallway.

"How could we possibly get there from here?" Brooks whispered.

"There's an airstrip right out that bunker door down the mountain about 100 yards. You didn't see it because you got here after dark. There are two supersonic jets that Latoya flies all the time from here to D.C. in four hours."

"Four hours? Not possible. And who's Latoya?"

After I explained it all, he perked up.

"It must be one of their new FX-40s," he said. "I've heard of 'em. Always wanted to see one."

"Well, now's your chance to actually fly in one. But your job will be to protect me when we get to Washington and get us back safely."

"Protect you from what?"

"From Muller, and Loren, and any of the others who don't want these experiments to succeed. I have to get to Washington to make sure my grandfather is safe and to retrieve the algorithm. Don't you get it?"

He lifted one eyebrow, "Impossible."

"Why?"

"A pilot, you, and me on a supersonic plane? There's no way. How many seats are on that plane?"

"Six."

Brooks' eyes opened wide. "You've got to be kidding. So, it's true. She must be huge."

"She's designed and designated solely to carry dignitaries, like cabinet members or congressional bigwigs, quickly anywhere around the globe. She's all computerized, one pilot, no navigator needed. She *is* huge. NOAA has two of them for this project alone. Both are here right now."

"Amazing. I heard some Americans saying she could shake plates out of kitchen cabinets."

I turned on my best, needy charm, "So, are we on? Will you help me? Latoya thinks very highly of you."

"I didn't need to hear that about some woman named Latoya. But yes, I'll help you."

– 42 –

BENNETT

BILL BENNETT WAS STRETCHED OUT over his Senate office desk, his hands cupped and covering his face. Perspiration covered his bald head.

Outside, several hundred protesters tossed Molotov cocktails and rocks at the capitol building police.

The protests and chanting had been going on for days. "You promised water! Where is our water? You promised water! Where is our water?"

Bennett eased the barrel of a revolver over his tongue and closed his eyes.

BANG!

– 43 –

Gail

THE TIME HAD COME FOR ME TO interact with Dr. Bernard Loren. I went to the far back corner of the simulator room where he was proudly explaining the simulator problems to Pete Gerritsen and his team of scientists. Pete and I made eye contact as I walked up behind the others. I listened.

I quickly realized how far off base Loren was with his "facts." Either he was lying on purpose, or he didn't know diddly about the simulator. I'd have to find and fix any problems myself, after hours. I noticed a small, spunky-looking female NOAA scientist in the group rolling her eyes on more than one occasion. I took note of it.

After the meeting broke off, I hung around until the young woman was alone. I approached her, studying her name tag.

"You're Gail!" I said. "I'm Grayson Fields. I designed the simulator."

"Really?" she said, surprised. "When Dr. Loren was explaining the problems to us, I saw you. I figured you were management."

"No, I'm not management." I invited her into one of the small conference rooms along the perimeter. She obliged. We walked to the corner, underneath the tiny camera in the ceiling, out of site from the lens.

"Gail, I'm going to let you in on something important. Please understand that you must keep this to yourself."

"I will."

"Dr. Loren was lying," I said. "And I know you know it, because I saw you roll your eyes more than a couple of times if I'm not mistaken. You didn't believe him either, did you?"

"No ma'am. What he said isn't possible."

Whew, I let out some air, thankful for her answer. I now know for certain she was not an admirer of Bernard Loren. "Tell me how you know his explanation contained some—shall we say—untruths. Then let's compare notes."

"Well, to start with," Gail said, "he said the iodide cannisters weren't large enough to bring the gauges up to 200 PSI in order to spray the chemical. That couldn't be right because I pressurized them myself when they first arrived, before anyone was down here."

"Excellent, I'm with you. So, we are on the same page. What else did he say that was wrong?"

"He explained that the load on the phosphorous drop should be 500 parts per million."

"But that's actually correct," I said. "How do you know he's wrong?"

"A week or so ago he used several of us to run a test one afternoon. I distinctly remember that we used 2,000 parts per million."

"Do you remember the date?"

"I need to look on my calendar. Give me a sec." She studied her watch. "Yes, here it is, September 12."

"Gail, that's the day I was on the plane that went down in the Bay. Hundreds of people died."

"Yes, I read where only twenty-four were picked up in life rafts. You mean you were one of them?"

"I was. Bernard Loren wanted me dead."

"My goodness. Why would he want that?"

"He's with Purity, not NOAA. They don't want this project to succeed. They don't want you and me to succeed. Tell me, why haven't you said something to NOAA officials about the storm that night?"

"Well, I just now put two and two together when you told me you were on that plane, and Dr. Loren's experiment that day. Besides, he doesn't like to be questioned. Everybody down here is afraid of him. I mean, his credentials and all. He's famous. We studied his textbooks."

"Gail, I'm glad you're here. I need your help. But here's the thing. Honestly, I'm not sure I can trust anyone in this facility. So, I have to ask you a couple of questions. Okay?"

"Okay, I guess so."

"First, may I look at your work card?"

She showed me the sealed card on her lanyard. I studied it. While handing it back I made a polite comment, "I see you've been here only two weeks. You were brought in much later than the others. Why is that?"

"Yes, I was working in NOAA's hydraulic section in Houston and my supervisor came to me. She must have thought my work was good, I guess—and asked if I'd like to do a temporary job, working with Dr. Loren. Naturally, I jumped on it."

"Perfect. And I suppose you've made several friends while you've been here?"

"Not really. People here aren't very friendly, and besides, I've been working two shifts."

"Gail, I know this may sound a little weird, but it will help me. Tell me the name of the person who discovered that liquid propane could be combined with silver iodides to manipulate the growth of clouds, and you'll be my trusted friend forever." I knew I was going too far with my questioning, but I wanted to make sure she wouldn't sabotage my plans.

She thought for a moment, then said, "Dr. Sorcroft? 2017? Is that right?"

"Close enough. It's Dr. Schwarzkopf. But are you familiar with his principle of the secondary effect?"

"Yes," she quickly responded, "It's when freezing water can release latent fusion heat into the cloud, causing them to grow into super large cells, too heavy to travel, so they dump all their water, right there, over the ocean. Right?"

"Perfect. You studied his work, I see."

"He's brilliant. More famous than Dr. Loren, for sure. I'd love to meet him some day."

"Well, I promise you'll be able to spend as much time with him as you'd like. As long as we can work together."

"Really? Of course. And I would—I would give anything to sit down and discuss some of his theories."

"Well, Gail, you'll have that opportunity. But, even better, you'll be able to witness it yourself; provided our next test proves positive."

"What are you going to do about Dr. Loren?"

"I don't know. But we should have the algorithm soon. So, will you help me?"

"Yes, ma'am, I will."

"First, don't ever call me ma'am. I'm barely older than you. Okay, look, here's what I need you to do. Go to the main terminal upstairs. Use my computer. I'll give you the password. I trust you. I need for you to go through the algorithm sequences they're using for all the variants. Somehow, they've written a virus formula into the sequencing. And I haven't been able to find it. Think you can find the culprit algorithm, and then let me know when you do? Once we take out the virus, we can install the new algorithm."

Gail leaned forward, her hands clasped in her lap, her eyebrows close enough together to register confusion.

"You can write a software program, can't you?" I asked.

"Oh, yes, yes I can, since third grade."

"Good. You should be able to do this. Listen closely. I believe Dr. Loren planted a virus in the system to sabotage our experiments. I wrote a simple formula to search all the algorithms for any odd character sequencing, but I couldn't find it. I'd like for someone else—you—to write your own version of a new search program and see if you can find it. I'm going to be at the simulator if you need me. You can do this, right?"

"Dr. Fields—sorry, that's your name, isn't it?""

"Yes. Dr. Grayson Fields. But, please, call me Grayson."

"Got it. One other thing. Dr. Loren mentioned that the sunlamp wasn't hot enough to create the evaporation needed. I saw the printouts from the experiment, and I know that's not my specialty, but I studied hydrology, and I swear, the ratios looked good enough to me."

"They are. I checked them before I came here. Now, think you can write the program and find the nasty culprit virus?"

"Yes ma'—I mean, you bet, I'm the little engine that could." And she took off.

– 44 –

Supersonic

LATOYA WAS MORE THAN PLEASED to fly me and Brooks to Washington. She just needed a good plan to make it happen.

"Grayson, I'll need to wait until there's an approved lift to D.C."

"What do you mean?"

"I mean I can't just run off and fly you two over there. Someone will either need to be picked up and brought over here, or there'll be someone from here headed back to D.C. I'll need to go there and give someone a hitch back here; that way, you don't have to fly outta here with one of their spies on board. Know what I mean?"

"I do. I'm just concerned about the time."

"Sister, it happens about every week or so. I can squeeze you in then. Like I say, NOAA is always flying people back and forth."

I actually said a prayer that we wouldn't be too late to take care of Grandfather.

Two days later Latoya found me in the cafeteria.

"I have a pick-up in D.C. who needs to come here to the facility. Some Senator. I can take you and the Brit over without a hitch. But you must realize that to get back, you'll be riding with a Senator. Is it worth the risk?"

"Depends. Who's the Senator?"

"I don't know. We'll find out."

Latoya was friendly with the bunker entrance guards, so they didn't ask anything when she, Brooks, and I casually walked outside and took the tram down to the airstrip.

As we approached the first plane, I heard Brooks blowing air through his mouth, and whispering, "My goodness, she's huge." We continued to walk toward the bird. She looked more like a black bullet; no discernable wings, just fins that protruded from the mid-section. Nose to fins made up 70 percent of her total length. I was nervous.

"I've already filed a flight plan," Latoya said. "We should be good to go when the tower gives us an okay." She pressed a device attached to her aviator shirt pocket, and I heard a whirring noise up above. Three black metal bars which were recessed into the plane came down to our level. One for Latoya, one for Brooks and one for me. Twenty feet above, three doors opened. "Brooks and Grayson," Latoya said, smiling at him, "take the lift up to your door. It'll take you up to your seats." Latoya rode up first. Brooks and I went up, side by side.

Once inside and in our seats, we buckled up, put on our headphones and I said, "Hello, can anyone hear me?"

Latoya blasted me through my headphones. "Honey, stay quiet. The tower could be alive. You can talk after we're wheels up."

"Lady Bird to tower, come in. Lady Bird to tower, come in."

"We got you, Lady Bird. Once you've punched out, you're clear to the sky. Have a good one."

Latoya went through a list of check clearances, while Brooks and I got settled into our seats. Brooks' face was all grins. My stomach was churning.

Latoya fired up the engines. We felt the vibrations. Latoya brought the big bird around to taxi position, moved her over to the start area and asked the tower for a go.

"You are good for go."

She pushed the stick forward, and my whole body melted into my seat. I remember feeling all organs being pressed against my back rib cage; my head was quickly pressed so hard against the head-rest, I couldn't even move it sideways no matter how hard I tried. Only my eyes could move toward Brooks. A big grin was still plastered across his face.

The next four hours were spent skipping across our earth's outermost atmosphere. "This is actually much more fuel efficient than flying those antiquated commercial monsters around the globe," Latoya let us know.

I settled in and let Latoya and Brooks get to know each other through their headphones. I thought there might be something blooming there. But Brooks was still grieving the loss of his wife and two daughters. Can't say as I blame him.

After waking from a short nap, I asked Latoya where we were. "Just coming over the Grand Banks. We should be in D.C. in forty-five. I called Opa three times once we'd made it that far. Still, no answer. We landed in four hours and fifteen minutes. I called Jesse.

"Dr. Fields, I've missed you."

"You missed me?" *Oh, my goodness he's not supposed to have feelings.*

"Jesse, I'm at Joint Base Andrews. Come to the front entrance and pick me up."

Latoya reported inside for her debriefing, while Brooks and I waited in the main lounge for Jesse.

The white Tesla soon drove up, but with no sign of life. I opened the front door expecting to say hello to Jesse. But Opa was in the back seat. "Opa!" I swung my arm at Jesse with a loud, "Jesse, so good to see ya." Then I went to the back door and slid in next to Grandfather. We hugged. "Why didn't you answer my calls?" I asked.

"My dear, my whole experiment station has been blown to smithereens. It's all gone." I could tell he was on the verge of tears.

"But Grayson, we have the algorithm."

Grandfather turned to Jesse in the driver's seat. "Jesse, tell my dear Grayson what new information you have."

"Dr. Fields, I have the algorithm that will create oceanic clouds and coax them over land."

We all cheered.

"Opa, what about Albert?" I asked.

"We don't know yet. For some reason, Jesse came and picked me up at work early today. Grayson, (his voice halted) Grayson, they blew up my simulator one hour later. One hour. It's gone, Grayson. I saw it on the news. There's nothing left."

Grandfather seemed feeble, weak, like he hadn't eaten in days. Or was it all the evil in the world dragging him down?

"Jesse and I were on our way back home when you called," he said. "We must check on Albert."

"We are. Right now," I said.

As we traveled through D.C., Brooks commented on the dry devastation, "I had no idea America had come to this, too. Surely God has been sending us all a message."

He was right. The trees, the bushes, everything was brown. I saw my first tumbleweed, not out West, but in our nation's capitol. It just rolled and bounced across Wisconsin Avenue as we passed the National Cathedral. *Our world is slowly withering into dust.*

We pulled through the majestic gates (creakier than ever) and stopped at the dry fountain. Grandfather opened the car door, but I stopped him. "Opa, let Brooks and me go inside first and make sure it's safe, okay? Please."

He twisted that pink mouth of his to one side and looked at me like I was the only person in the world that mattered. "Okay," he whispered.

Brooks and I entered the front foyer, saw nothing unusual and proceeded to search the first floor, calling out for Albert. "Albert, where are you? Albert!"

We found him on the second-floor landing. He was dead. We didn't know how, or by whom.

In the distance we could hear sirens; faint at first but picking up steam. "If we take him with us," Brooks whispered, "we'll never make it out alive before they get here. We have to go now. They'll take care of him. They will."

"But I hate to just leave him lying here."

Brooks grabbed my arm, "Grayson he's dead. Go with me now or they'll put you in jail for the rest of your life. C'mon, now!"

We quickly ran down the stairs, out the front and jumped in the Tesla. "Go, Go, Go," Brooks yelled at Jesse. We peeled out so fast I remembered the G-forces from Latoya's rocket.

We made it to the first curve in the street when I turned to look back and saw a legion of police cars enter the front gates.

"Opa, we need to get back to the airport. To Andrews. We have a plane waiting for us. For all of us."

Grandfather was indignant, "No, no, no, we are going to turn around and get Alfred."

"Opa, none of us is safe if we go back. I found him, Opa. He's gone. He's dead. Murdered. There was nothing we could do."

"Murdered? Oh—I—I can't imagine. Why would they—"

"Opa, we're going to the air base and that's final."

He twisted his head to one side and didn't take his eyes off me. "My child, then I at least need to see my simulator."

"But you said it was gone, blown up!"

"Haaa," he sighed, "I've spent so many years in there. Please, let me see what it looks like."

"Okay, Opa," but just for a few minutes. Our plane is waiting for us." I turned to Jesse, "Take us to Georgetown, to Grandfather's simulator."

We arrived twenty minutes later. Police, SWAT personnel, even the FBI were on the scene. All I could see was police tape stretched far outside the entrance to what used to be the old Hoya gym and Grandfather's simulator. Bomb debris was everywhere.

I looked over at Opa. He had dropped his head onto his chest. His eyes became like those of a man on his last breath. I actually said a little prayer for him.

Not five seconds later, he perked up.

"Jesse, take us to the airport and fast," he said. "And who is this man?"

"This is Brooks. He's a missionary."

"A missionary? Child, what have you been doing?"

"I'll explain it all later, Opa." I turned to Jesse. "Jesse do not give that algorithm to anyone, except me. And not before I ask for it at the facility. Do you understand?"

"Yes, Dr. Fields, I understand. What is the facility?"

"Jesse, you're coming with us to India."

"India? Yes, I know India. I know about all the countries in the world. My prior knowledge was given to me long before I came to work for Dr. Schwarzkopf. I believe there are bad men at this facility. I read where some—"

"Jesse, stop. Thank you, that will do. And yes, there are bad men there. But Jesse, we will be okay." I was too keyed up to become emotional.

Opa grabbed my arm and held on. "We're all going to India? When?"

– 45 –

JESSE

OPA, JESSE, BROOKS, AND I arrived at Andrews on the tarmac next to the only FX-40 on the base. We left the Tesla and walked toward the big bird.

I looked over at Jesse. He had nothing to say as we continued to walk. Latoya was already up in her pilot's seat, the engines warming up.

The hydraulic bar lifts came down and we rode up two at a time. After we settled in and buckled up, I told Latoya, "Ready for take-off, Captain."

"Not so fast. The Senator hasn't arrived yet."

We waited another thirty or so minutes until a black Cadillac drove up. Senator Chambers soon made her entry inside the cabin with an announcement, "Oh my, I didn't know we'd have company, Latoya. Let's see, we have Dr. Fields, 'Hello,' and Julius Schwarzkopf, it's good to see you Julius. Who might these other travelers be?"

"I'm so glad you asked," I said. "This is Brooks, he's a missionary from India, and this is Jesse, he's our personal man-made. And let me add—" I started to say as Latoya broke in on the cabin speakers,

"Buckle up. Wheels up in thirty seconds." We were already taxiing.

"Senator Chambers, let me add that we're all here on a mission and it's not the same mission we suspect you're on."

"Well dear, tell me your mission and I'll tell—"

At that moment, we were all pushed back in our seats and couldn't move. We shot off like a rocket, slammed against our seats so fast we were unable to move. The jet engines drowned out all voices except what would come through our headphones, and even then, all I could hear were human moans.

It wasn't long before we were in and out of weightlessness, skipping along the Kármán line, 62 miles above the earth.

I didn't bother to turn around and face Chambers. I just spoke into my mask. "Senator Chambers, our mission is to make ocean clouds migrate and drop their fresh water over land. I'm pretty sure your mission is to make sure that doesn't happen."

"Not exactly," she said. "My mission is the same as yours."

I turned to Opa to make sure we were all on the same headphone frequency. "Did you hear her, Opa?"

He nodded but his quizzical eyes said he wasn't believing her. I turned to face her in the back seat. "Good, then we'll all stick together, right?"

"Dr. Fields, I am a U.S. Senator. I go where I please."

"Of course, you do. There's only one catch. We believe you work for Purity, and besides, we're glad we have you on this flight because I have some questions for you. For example, you know why my parents went down in that helicopter don't you?"

"Wrong," she said. "Your grandfather here asked me to investigate the NTSB's findings. I couldn't find anything wrong with their report. It was pilot error."

Latoya's voice cracked through our headphones. "Not true, Senator. I knew the pilot. He didn't make mistakes. He'd never had one demerit in his entire career. My sources in the bay tell me it was caused by mechanical problems or by some explosion. It would have been nearly impossible for the mechanics to miss something that could bring it down, including a bomb on board. My team concluded it was a surface to air missile. A child could've hit that helo with a handheld launcher. The report had to be falsified."

"Don't be so sure of yourself, Latoya," Chambers shot back. "You didn't see the report. I did."

"And therein lies the problem," Opa said.

"Ladies and gentlemen, and Senator Chambers, I suggest we all settle down and you get some rest while you leave the driving to me. We'll be landing at Darjeeling in three hours, fifty minutes."

I was too tired to think about talking anymore, so I increased my oxygen to max and fell asleep.

After landing, we took the tram up to the bunker entrance. We were allowed inside immediately. On the way in, I stopped to hug Latoya and thank her.

"Anytime, girl."

Brooks stopped. "Latoya, thank you for the thrill ride of my life."

"Well, you're very welcome, Brooks. But I believe you owe me a cup of coffee." I was tickled by Latoya's forwardness toward Brooks.

Brooks froze. He didn't know what to say. "I—I suppose okay. When?" His hesitancy was surely brought on by the pain of his continued grief.

I left them and walked on ahead to catch up with Senator Chambers. She was standing in the lobby, mesmerized by its sheer size. "It's so unlike anything I've ever seen."

We were inside the facility's entrance foyer, eighty feet to the ceiling, IBM computer banks stacked and racked the length of the room.

"If you think this is something, Senator, let me take you to the simulator," I said. Grandfather caught up to us. "Opa, you come too. We can get you both checked into quarters later. You need to see this first." As we started for the elevators Brooks caught up but said he was headed down to see Rinku and Tahir. I guessed that Latoya was back in the pilot section filing her report.

I could never have imagined what would happen to Rinku next.

– 46 –

The Secret Room

IN THE INFIRMARY, BROOKS, RINKU and a nurse were standing beside Tahir's bed. Once the nurse finished explaining Tahir's prognosis—which was good—Rinku whispered in Brooks' ear, "There's a secret security room. I saw it."

"Listen to me, Rinku. Do not go near any bloody security room. Just stay here."

There was a long silence as Brooks thanked the nurse and turned his attention to Tahir.

"I have to pee," said Rinku.

Brooks glanced to see Rinku leave the room.

Rinku never looked back. Neither did he ever intend to use the restroom. He walked up a level, and down a hallway to just outside the MAINTENANCE room. Earlier, he had seen Muller and Loren entering that same room. Rinku slowly twisted the doorknob. It turned. He was surprised. His heart rate jumped thirty beats faster. He remained quiet and carefully pushed the door open one inch. He listened. Loren was talking on the phone, oblivious to Rinku.

"Call it want you want," Loren said "she's got to go. For good. If she finds the algorithm, and this thing works, we're screwed. Look, just make sure she doesn't get down there and try to conduct her own experiments. Understood?"

Rinku was leaning on the door and it opened another inch or two. Loren noticed.

"Whoever the hell is out there you might as well come on in. Steve, is that you?"

Rinku stepped in and saw all the monitors, dozens of them embedded up high in a curved wall. Rinku could have stepped straight forward to where Loren sat, but instead he eased over to the monitor wall and kept walking farther into the room. He alternated glancing up at the monitors and back over at Loren. It was dark inside, except for the glow from the monitors and desk lamps.

Rinku was scared. He gritted his teeth to keep his cool and not bolt back out the door.

"You little twit," Loren said with a guttural sound of dark evil Rinku had never experienced. Now Rinku wished he hadn't walked in. "You're scared out of your mind, aren't you?"

Shaky, Rinku said, "After what I've been through, do you think I'm scared, sir?"

"Well, listen to you. Even a little pissant can have nice manners."

"Yes sir, but I don't think you're very nice."

"Tell me—what the crap do *you* think is going on around here?"

"Sir," said Rinku, "I think—I think you are watching everything that goes on here (Rinku turned to look up at the monitors). And I think you and Mr. Muller are making people sick with your tainted water. I tasted it. I think you are sabo—sab—o—"

"Sabotaging, is that what you're trying to say? You really are a little twit. Come on, say it with me. *Sab-o-tage*. Not even close you little prick. I'm not sabotaging anything, except you!"

Rinku remembered his phone. On it was an alert key to reach Brooks in an emergency. He pressed the key twice.

Loren scrambled over the console top, jumped on Rinku, and quickly pinned him to the floor with muscle and force. But Rinku was agile enough to keep Loren from getting a good grip.

Loren finally got him in a firm headlock. Rinku was helpless, gasping for air; he started to turn purple when his blurry vision caught site of Brooks entering the room.

Brooks was on Loren in seconds, knocking him off Rinku and putting Loren in a military death chokehold. Loren didn't stand a chance. Brooks' massive arms were already glistening with sweat. His eyes, those of a soldier who tastes the kill.

Rinku scrambled to his feet, still holding his throat, "Father, do not kill him. Please. What did you teach me? Vengeance is *his*. Not ours," he shouted.

Brooks' eyes were still in kill mode, sweat dripping down his face. Loren was fading into unconsciousness, his eyes on Rinku.

"Father! Stop!" shouted Rinku.

Suddenly, Brooks released Loren and tossed him to the side like a rag doll. Brooks turned to Rinku.

"Son, find some duct tape or something, and some rope. Now!"

Rinku hustled out the door.

Ten minutes later, he walked back in, sweating, breathing heavy, holding a roll of duct tape and rope.

Brooks tied Loren up and covered his mouth with the tape, threw him over his shoulder and walked out of the security room. Rinku asked, "Where are you taking him?"

"As far down in this facility as I can get him. To the bottom."

– 47 –

The Algorithm

IN THE MAIN COMPUTER ROOM, I made a beeline for Gail.

"You did it, Gail! You found the culprit virus. Now, we have the algorithm and that means we have one more mission to complete. We need to make rain, and we need to coax it to move over the land on that simulator. Are you ready?"

She pumped her fist and tapped me on the edge of my shoulder, "I'm ready, Dr. Fields."

I then introduced Gail to Jesse.

"Jesse will give you the algorithm. I'll meet you in the simulator room after you've set it up on the computer. I have someone I need to see first."

Brooks and I took two NOAA security guards and walked into the MAINTENANCE room to find Muller. When I saw Senator Chambers at the console next to Muller, I knew we had caught two dirty birds in one fell swoop.

"Have you been looking for Dr. Loren?" I asked.

"You know where he is?" Muller asked.

"He's in the basement," Brooks said. "He tried to kill Rinku."

I chimed in, "Now, would you like to see what we can do when we have the guts and the smarts to make ocean clouds move over land?"

"You can't do it, Grayson. It's impossible. Even you know that."

"We won, Steve. Yes, you, the great Steve Muller. You are about to see what your future looks like. And it's not a pretty picture."

"Guards, would you escort Mr. Muller and Mrs. Chambers to the—the, what's that name again, Brooks, I forget?"

"The brig," he said, smiling.

"Yes, that's it."

Muller stared at Brooks, deciding he was no match for Brooks and the two guards. "This is far from over, Grayson. The experiments will fail. And then we'll be switching places," as the guards handcuffed him.

In return, I had only one thing to say: "Steve, when you ignore the inevitable, the inevitable will surely bite you in your ignorant ass."

Neither was Senator Chambers willing to capitulate. "You little bitch, do you think you have any proof I've done anything wrong here?"

"In fact, I do. Seems your colleague, Senator Bennett left a note before he committed suicide. Explains your involvement from day one. And Steve, the Department of Justice will not be looking favorably at the virus you and

Loren installed on the computers. We found it, and it's gone. Now, both of you will be picked up by the FBI at Andrews. So, if you don't mind, I'm going down to Level 5 and make an ocean cloud migrate over land and replenish the very thing that will bring life back to our world. It may even put Purity Worldwide out of business, who knows."

The whole time I was saying this my voice was having to compete with the agitated, screaming voices of Margaret Chambers and Steve Muller. I don't know what they said. I just know they weren't happy.

Muller put his arm around her. "Don't worry, Margaret, I'll have us both out of this once we get to the states."

"You better," she said, her eyes were slits and her lips as taut as a prune.

Muller turned to me, "Grayson," he said, shaking his head and blinking rapidly, "you will surely regret this. You have no idea what I'm capable of." His face grew more contorted and twisted. "You will need to hide somewhere for the rest of your damn life. I will find—"

"Steve," I interrupted. "I can assure you of one thing: You're going to be in prison for several years." I began to walk off, "And I'm sure you'll meet some beefy new friends while you're there."

The last I saw of Steve Muller and Margaret Chambers was of their backs, their heads hanging down, their arms behind them in handcuffs.

As Brooks went with the guards to the brig, he said, "I'll join you later after I see how Tahir is doing."

I rode the elevator down five stories to the simulator.

Gail, Opa, Jesse, Gerritsen, Latoya, and a dozen scientists were inside the enormous glass bubble waiting for me.

I walked to the iris scanner, scanned my eyes, and stepped through the hissing glass doors.

Just inside was a new sound studio of sorts with a technician ready to make any announcements to everyone inside the simulator. I took the microphone and made my own announcement.

"Ladies and gentlemen, NOAA scientists, everyone in this room, we have work to do. Henceforth, Mr. Muller and his team of scientists are no longer needed for our experiments. NOAA is in control." A big cheer went up by everyone inside the simulator.

"Now I need Gail. Gail, where are you? Raise your hand so I can see you."

She was at the other end, 100 yards away, waving. I could barely see her.

When she reached me, she was out of breath. "Gail, didn't I promise you you'd be able to make history here in this room? I want you and Jesse to go upstairs to the computer room, where you'll enter the new algorithm. Then, come back here. We'll all see if this new algorithm is the real thing."

– 48 –

Are You the One to Make it Rain?

I WALKED FROM ONE END OF THE simulator to midway up the right side. I climbed up a tower ladder surrounded with a railing, a table with computers, and another microphone. I was looking out over a platform the size of a domed NFL football arena. I took the mic and addressed everyone again, "Can everyone hear me?"

From one end of the football-size simulator to the other came voices confirming everyone could hear.

"First, we're going to get the lab batch containers cleaned and load the new algorithm when it comes down—which should be any minute. So, I'd like for you, and you (I pointed to a couple of scientists with NOAA lab coats) to clean the tips on the atomizer nozzles. You, I need for you to push the 'go' button on the mini-drones." I looked around for one more volunteer. "Would you be so kind as to check on Gail upstairs and get a progress report on the algorithm?"

She jumped at the chance.

There was not much more I needed to prepare for. I asked each scientist to give me an update on their equipment. Each gave a thumbs up.

The computer behind me buzzed. I jerked around to see a call coming from the main computer room upstairs. *That's gotta be Gail. I prayed she and Jesse had successfully loaded the algorithm.*

```
All is well. Sending algo.
```

I turned to look down at another NOAA scientist. "Now it's your turn! Get a lab batch ASAP."

The entire sequence of events would be handled by the computer. The algorithm drives the process, the timing of each event, and the amount of chemicals to be used. The scientists at each station are the back-ups.

It was time for me to begin the show. "Don your sunglasses." I turned and pressed the COMM S key on my computer to start the first act of the show. The sun.

All lights in the cavern and inside our dome went dark. The sunlamp, at the top of the dome, became the only light source. We instantly felt the heat. And I felt the goosebumps but knew all too well this sun was about as dangerous as the real thing. Our scientists knew not to stare at the lamp for any extended time.

The bright light 100 feet above us was powering up to max sun-simulation.

The room looked eerily like the real earth and ocean with a blazing sun in the afternoon. Shadows appeared around all the miniature objects.

The sun lamp kept heating the ocean's surface. The temperature everywhere began to rise, by the hour. Everyone began to wipe sweat from their foreheads.

We monitored progress on the computer.

I would often turn and see what it looked like in this realistic miniature world called a simulator. The cloud forming above is always a thrill of mine. Most of these scientists had never seen such a thing. "Oh, my goodness" spread from one scientist to the other.

I asked each scientist manning a station around the platform to nod if their systems were a go. One by one, they all nodded.

I was so nervous I could only nod my affirmative response to each one as they gave me the green light.

Enough evaporation had already occurred to form a decent size cloud over the miniature Bay of Bengal. I could never grow nonchalant seeing a spectacle such as this.

I asked Opa to climb up on the tower with me. He needed to be the one to guide this special moment. Gerritsen helped him up and joined us.

I handed Opa the microphone. "Here. This is for you. You, of all people, should be the one giving them the countdown. Here, go ahead."

"But it's all computerized," he said, his eyes already misty from getting this far with his life's work.

"Yea, but you can still press the key on this computer to start the program." He looked at me with those intense eyes and wrinkly, ever present crow's feet, and a wide grateful smile that said, *I love you.*

Taking the mic, he looked out over the simulator and all the scientists stationed at various stations around the platform.

"I want to thank all of you. Thank you. That is all I have to say." He handed the mic back to me. I twisted my head and gave him my best coy smile, "Oh, Grandfather, you can do better than that. Here, try again."

"I'm just a bit overwhelmed. Okay." Holding the mic up again, "All scientists stand in place. Three, two, one, gehen . . . go!"

I grabbed Opa's hand and led him to the S2 key.

This key kicked off the rest of the computerized sequences. The needles protruding from underneath the platform over the Bay began to release the white phosphorus.

We all waited. The cloud practically exploded into a mean giant, growing darker and menacing by the minute. Electrical charges popped inside.

Six miniature drones from the Bay end of the simulator came out of hiding and slowly buzzed over the simulator, leaving a chemical spray. They headed toward land. As they reached the coastline, they crossed paths with drones discharging positive charged ions coming from the land.

This was the moment so crucial.

We watched in amazement. Five minutes or so after the drones dropped off the platform, we noticed the cloud moving. Moving toward land.

"Go. Go. Yes, keeeep going. Keeeep going," I whispered.

I could hear scientists inside the simulator saying "Keeeep going."

The cloud reached the land's coastline. We all jumped and screamed, "Yes, yes!"

"Finally," I screamed. I looked over at Opa. My eyes were full of held-back tears. His eyes had misted over. He just shook his head in disbelief and joy.

"All these years, and finally, I get to see it," he said.

Another valve with atomized silver iodide spewed from the nozzles above into the cloud's center. In a matter of minutes, the cloud grew darker, heavier. The crackles in electrical charges inside the cloud sent several scientists underneath the platform.

I wasn't sure if any of us were safe being inside this enclosed space with a cloud that now dwarfed the entire width of the platform and was so dark, the sunlamp was no longer visible.

Lightening striking the platform became an issue. Especially for us, ten feet above the platform on a metal stand. *Why didn't we think of possible lightening like this?*

Rinku came up below us from outside and knocked on the glass wall. I shook my head and mouthed, "Can't open door." He understood and watched the action from outside our eco-system.

The cloud kept growing and traveling farther inland. Lightning bolts popped the land. Everybody jumped back, startled.

I turned to the computer screen behind me, looking for any cloud rotation indicating tornado activity. *None.* I was thankful.

"Go. Go. Go baby," Gerritsen said louder and louder.

Opa said, "Keep going, my grand damsel, keep going."

Our cloud continued moving over land. Soon the bottom fell out. The cloud continued inland and eventually covered the entire eastern provinces of India right in front of us on the scale model platform. Everyone celebrated like we'd won the lottery. Looking in from the outside, Rinku jumped up and down, pumping his fist in joy.

I grabbed the mic from Opa. "That's it. That's it! The algorithm works. We're ready to put fifty thousand pounds in production asap. Shouldn't take but a couple hours." I spotted Lisa Chow, NOAA's chief architect, "Lisa, will you please peel the algorithm off the computer and get the chemicals ready for the drones?"

Gerritsen called the pilot's center and told them the payload would be on the tarmac in two hours.

I sent someone to let Rinku in.

Minutes later: "You did it! You made it rain!" he cried, hugging me like he meant it.

"I know, Rinku, I know. Where is Brooks? He should see this. We're preparing for the real drones to take the chemicals and make it rain for real. Out there!" I pointed to the outside.

"My father is still with Tahir, praying in the infirmary."

"Go get him. He needs to see this." I turned to see Gail below our tower. I yelled, "Gail, you did it! We did it. Get up here."

As Gail was climbing up, I heard Opa's phone ring. Three times.

"Opa, your phone, it's ringing. It could be important."

"Hello!" he said, his voice agitated from being interrupted.

He covered the phone and turned to me. "It's John. He's calling from the Oval Office.

"Mr. President, you have better things to do than call me."

Mr. President? What the? I leaned in close to hear. I recognized the president's voice.

"Julius, when can you get back here? There's going to be some big celebrations in your honor, my friend. I even have my team arranging a parade here in Washington."

"John, we've only proven the algorithm in this simulator. We need to get it out in the field and make sure it'll scale up and make real rain."

"Julius! What are you talking about? We don't know what you did over there but it's raining cats and dogs in several places around the globe. Hell, it's raining here in Washington! I got word that Paris is getting rain. And get this, Africa and India, even where you are."

"Where I am? Here? You're saying it's raining somewhere in India?"

"That's exactly what I'm saying, Julius. Now get your old butt back here so we can give you the proper medals you deserve. You've made history, my friend, and I'll personally be nominating you for a Nobel. Hell, for two Nobels."

"Thank you," Opa said, and hung up, confused, and pale.

"Did I hear him say something about rain in Washington? What did he mean? What was he talking about, Opa?"

Opa couldn't stop shaking his head. "He thinks we sent planes out all over the globe and made it rain. He said it was raining in Paris, and . . . and, I can't remember where else. Oh, he said it was raining in India. In India, Grayson. In India. I wasn't able to tell him we had nothing to do with it."

My mouth dropped open. We climbed down from the tower and ran—as fast as Opa could—for the elevator.

Meanwhile, in D.C., the Speaker of the Senate was at the dais. The chambers were packed. She pushed the button that sounded the electronic gavel, "BANG!" to make an announcement. "Ladies and gentlemen, the vote on S5482, sponsored by Senator Bennett—our recently-departed friend and colleague—and by Senator Thompson, was approved by a vote of 52 to 47. Our national water supply will remain under private industry control, with new oversight from the Secretary of Water Sustainability."

Gerritsen and Gail joined Opa and me in the main foyer, along with a dozen NOAA scientists. We bolted out of the bunker to see if it could possibly be raining here.

When we reached the outside it wasn't just raining, it was pouring. The bottom had fallen out. We all held our heads back and let the rain drench our faces, our hair, our clothes. We couldn't stop laughing. The rain's collision with the concrete under our feet made such a noise we could barely hear each other's expressions of wonderment.

And the smell of the rain; oh, I'll never forget how fresh it made me feel. Pure. Alive again. The hair on my arms and neck stood at attention. No amount of water could keep them down.

Gerritsen shouted, "Did someone figure this out before us? What the hell happened?"

"I'm as astounded as you," I said.

"We didn't have time to get a single drone up. And this? I don't know what to think any more, Grayson. Did you do something I don't know about?"

I turned to see Brooks and Rinku coming out of the bunker, bringing Tahir in a wheelchair. "Here comes your answer now." I could only stare at Tahir, Brooks, and Rinku, and wonder about this God of theirs.

The cloud-seeding spray planes were only a hundred yards away on the tarmac below. The engines shut down, one by one.

Latoya came out. She couldn't believe it either. I watched her cozy up to Brooks. They talked and laughed and held their faces up to the sky, the rain still drenching them and every one of us. Rinku came over and held my hand. He was my little brother, now.

Opa came out and put his arm around me. The three of us huddled together. Brooks gave me a big smile. I smiled back, and nodded several times, pointing to Latoya the whole time. He turned his head to one side with a half shake, signifying to me he still wasn't quite ready.

I looked around for Jesse. He was standing under the entrance door overhang. I had forgotten how much he despises the rain. It would screw up his circuits.

Unbeknownst to me at the time, ground water from the torrential rain was rushing down a slope behind the building's lowest level. Water was pooling up in the basement.

Tied up and laying on his side, Dr. Bernard Loren was squirming and moaning, only able to focus on the muddy water rising around him.

— THE END —

ENDNOTES

[1] *Blue Covenant*, Maude Barlow, The New Press, pg. 54

[2] *IBID*, pg. 55

[3] *Oxford Atlas of the World, Deluxe Edition*, Oxford University Press, pg. 56

[4] *Blue Covenant*, Maude Barlow, The New Press, pg. 5

[5] World of Rivers, Supplement to National Geographic, April 2010

[6] *Blue Covenant*, Maude Barlow, The New Press, pg. 115

[7] *IBID*, pg. 3

APPENDIX

FACTS ABOUT WATER

What *is* water? Water is a substance with the chemical formula H2O—one oxygen and two hydrogen atoms combined. Pure water is tasteless and odorless.

Water covers 70.9% of the earth's surface. Ocean water accounts for 97.4% of earth's water; but of course, it's saltwater and not potable.

The other 2.6% is fresh water, which includes swamp water, rivers, lakes and the like. Of that 2.6% of fresh water, 76.6% of it is locked in polar caps and glaciers. Another 22.7% is contained in underground aquifers.

That leaves us with 0.5% "active" water, from our lakes, rivers, the atmosphere, and soil moisture. That's it—one half of one percent. Lakes account for 52% of this 0.5% of total earth water. Rivers, only 3.5%.

*Atlas of the World, Oxford University Press, 2005.

The Great Lakes alone hold 84% of North America's surface fresh water and 21% of the *world's* supply. All of them are under duress from either algae or shrinking shores.

U.S. Army Corps of Engineers, May, 2013

Now consider that about 70% of "available" water is consumed by agriculture. One report in 2009, estimated that, in developing countries, water demand will exceed supply by 50%. *

*Atlas of the World, Oxford University Press, 2005.

___ "In 2003, two-fifths of the world lacked access to proper sanitation, which has led to massive outbreaks of waterborne diseases." Blue Covenant, Maude Barlow, pg 3, from the World Health Organization (WHO).

___ "Every eight seconds, in 2007, a child died from drinking dirty water." (Another article counted the death rate at every fifteen seconds.) Blue Covenant, pg 3, from WHO.

___ "Half of the world's hospital beds are being occupied by preventable waterborne diseases." Blue Covenant, pg 3, from WHO.

___ Americans use 100 gallons of water each and every day, at home. The world's poorest live on 5 gallons. Women in developing countries walk an average of 4 miles to get water in containers of all kinds that weigh 10 to 20 pounds. Forty-six percent of people do not have water piped into their homes. ~~from various sources

___ "The parts of the Earth, as far back as 2002, running out of drinkable water (known as "hot stains") included Northern China, large areas of Asia and Africa, the Middle East, Australia, the Midwestern U.S. and sections of South America and Mexico." Blue Covenant, pg 3, from WHO.

___ In America, one article was high-lighted with this fact: "Forty percent of rivers and streams are too dangerous for fishing, swimming, or drinking; as are forty-six percent of lakes due to massive toxic runoff from industrial farms, livestock operations, and over one billion pounds of industrial strength weed killer per year." Blue Covenant, pg 5, from WHO.

___ "In 2007, Lake Superior, the world's largest freshwater lake, dropped to its lowest level in eighty years, and has receded more than fifteen meters (49 feet) from the shoreline." Blue Covenant, pg 4, from WHO.

___ In India, "Seventy-five percent of the rivers and lakes are so polluted they cannot be used for drinking or bathing. Over two million Indian children under the age of five die every year from dirty water." "IBID."

___ "In China, eighty percent of the major rivers are so degraded they no longer support aquatic life; an astounding 90 percent of all groundwater systems under the cities are contaminated." "IBID."

___ "In Latin America and the Caribbean, more than 100 million people do not have safe drinking water." "IBID."

___ "In Africa, more than one-third of the people currently lacks access to safe drinking water. All of Africa's 677 lakes have deteriorated." "IBID."

___ WATER WITHDRAWAL

"The amount of water withdrawal as a percentage of the total water available is forecast to rise substantially by 2025. As population growth continues, the number of people affected by water stress and scarcity will increase significantly. During the past century, global water withdrawals have increased six-fold."

Global Water Intelligence, March 2010.

___ RISE IN WATER DEMAND

"Even with population growth being the major contributor to elevated global water-withdrawal levels, the rise in demand for water has outpaced population growth by a factor of two."

Global Water Intelligence, March 2010.

___ DROUGHT

"Steadily increasing temperatures associated with climate change and widespread exploitation of water resources across the globe have made droughts a recurring and growing threat. A National Center for Atmospheric Research (NCAR) study concluded that the U.S. and many other populous countries face an increased threat of severe drought in the coming decades."

Dai, Aiguo. "Drought Under Global Warming: A Review," Wiley Interdisciplinary Reviews: Climate Change.

___ PRIVATE FINANCING

"Private water financing was less popular in the 2000s, but this trend is expected to be reversed as municipalities attempt to rein in spending and balance their budgets. Private-sector participation in the water industry should benefit from increased demand for advanced water and wastewater technologies."

Global Water Intelligence, March 2010.

___ SNOW COVER

"In the next several decades, the ice albedo feedback--whereby melting snow exposes more dark ground, which in turn absorbs heat and causes more snow to melt--will accelerate the rate of Arctic sea-ice melt. Projected increases in global air temperatures will diminish the extent of the snow cover and induce premature snow melt; this reduction in overall snow cover will itself exacerbate global warming." Overland, James E., John E. Walsh, and Muyin Wang. "Why Are Ice and Snow Changing?"

Global Outlook for Ice and Snow (2007): 29-38.

___ URBANIZATION

"Urbanization leads to increased water demand, for both household needs and services. Household needs, which include activities such as flushing a toilet, watering flowers, or washing a car, increase daily per capita water needs by 80 to 250 liters. Service institutions, such as hospitals, restaurants, and hotels, are also major consumers of water. Urbanization can cause the demand for water to increase five-fold beyond the "basic water requirement." This increase does not include water used in power generation or other industrial activities that typically accompany urbanization."

Source: Alexander Zehnder et al., "Water Issues: the Need for Action at Different Levels," Aquatic Sciences, 2003.

MORE RESEARCH
FOR THE CURIOUS

<u>How It Happened</u>

In 1000 A.D., about 300 million people lived on our planet. By the late 18th Century, the Industrial Revolution created a population explosion and 1 billion tilled the soil, slaved in industrial plants, and continued to clear the forests. Seventy years later (1920), the population had doubled to 2 billion; and doubled again to 4 billion by 1975. By 2000, the earth was failing to feed all 6 billion inhabitants. Deforestation had created an ephemeral oasis of farmland, but erosion quickly re-claimed it to desert land. (1 billion lacked safe drinking water, and 2.6 billion lacked basic sanitation. *)

In 2025 half of the world's population will face serious water shortages. Over 300 major rivers cross frontiers and country borders, leading to disputes, conflict, and wars. But the population growth will march on, and by 2032, the world will likely grow to over 9 billion people. When, not if, a multi-year drought occurs, the world population will most likely shrink by two to three billion over a five-year period due to starvation at a faster rate than the growth occurred.

Land-locked countries and cities without a port to receive treated water will be in the worst condition. But even countries that once had life-sustaining rivers running through them, will lose their lifeline. Aquifers around the globe are already being depleted. Population explosions only put more pressure on water and food resources. Since 2002, these places have been officially known as 'Hot Stains.'

*Atlas of the World, Oxford University Press, 2005.

"The world is running out of available, clean fresh water at an exponentially dangerous rate just as the population of the world is set to increase again. It is like a comet poised to hit the Earth."

Blue Covenant, Maude Barlow, The New Press, 2007

THINGS TO COME?

Niger, Africa

If Mexico City was the poster child of land-locked cities, Niger, Africa, is the worst land-locked country. With a population of nearly 16 million in 2010, Niger will grow to 23 million by 2030. It's already a veritable wasteland. So awful has Niger become, that its population could dwindle to only 15 million by 2035, the vacancies caused by out-migration and starvation.

Buenos Aires, Argentina

With the renowned river Parana, once flowing from the Brazilian rain forests, now nearly empty, the Argentinians living in Buenos Aires were, in just three short years, thrown into shock. Water-ladened tankers entered the port daily, to be off-loaded onto tanker trucks. The trucks were being hi-jacked by hoodlums, with no compunction about shooting the drivers and guards. Government buildings had been taken over by the revolt. Man or woman, if they looked wealthy, they were targets. Victim's IDs led the water takers to the deads' homes, in search of more water.

Geidam, Nigeria

Another "City on a River." Geidam was situated on the north side of the Yobe River. But no matter now. The Yobe, and all mid-size rivers, had practically vanished. Dried up. After five years of tribal warfare over water rights, people had lost their appetite for fighting. Before annihilating each other, they came to their senses and banded together. To dig. And dig, and dig. For water. Every day. But it was too late. No longer was there an aquifer along the Yobe. By 2035, Geidam had vanished.

India

In 2032 India took over China in the population race, growing to 1 billion, 5 hundred million. But, during the three-year drought starting in 2032, India's population was reversing at as rapid a pace as it had grown. Especially hard hit was central India all the way north to Delhi. In the early 21st Century, Delhi was the 12th largest city in the world, with over 16 million inhabitants. By 2035, it had ballooned to become the third largest—22 million, but dying off, like all the hot growth cities in the world, faster than their previous growth rate.

MORE

Researchers used cloud observations from 1997 to 2009 collected at the ARM SGP observatory to generate an automated algorithm that classifies clouds into seven types: low clouds, congestus, deep convection, altocumulus, altostratus, cirrostratus/anvil, and cirrus. The researchers based this classification on the physical qualities of cloud top, cloud base, and physical thickness of cloud layers measured with millimeter-wavelength cloud radar and micro pulse lidar.

Additionally, the researchers developed another algorithm to identify fair-weather shallow cumulus events using cloud fraction information collected from 2000 to 2008 with a total sky imager and ceilometer. The events identified automatically agreed closely with fair-weather shallow cumulus events identified manually. The automated analysis only missed six cases out of 70 possible events during the spring to summer seasons (May–August).

Automated identification of cloud types at the ARM Southern Great Plains site

Submitter

Gustafson, William I. — Pacific Northwest National Laboratory

Riihimaki, Laura Dian — CIRES | NOAA ESRL GMD

Area of research

Cloud Distributions/Characterizations

Summary

Seeding of tropical cumulus clouds, and indeed any cloud, requires that they contain supercooled water--that is, liquid water colder than zero Celsius. Introduction of a substance, such as silver iodide, that has a crystalline structure similar to that of ice will induce freezing. In mid-latitude clouds, the usual seeding strategy has been based upon the vapor pressure being lower over water than over ice. When ice particles form in supercooled clouds, they grow at the expense of liquid droplets and become heavy enough to fall as rain from clouds that otherwise would produce none.

Other novels by author:

The Rector

The Actress

DAVID, Vol 1

DAVID, Vol 2

The Parchman Preacher

CPSIA information can be obtained
at www.ICGtesting.com
Printed in the USA
LVHW042007150521
686878LV00002B/1/J

9 780997 655605